journeys
fuelled by ideas

Made in Brunel
Brunel University, Uxbridge, UB8 3PH
+44 (0) 1895 267 776
www.madeinbrunel.com

Art direction by Edwin Foote

Editors:
Vincent Ashikordi
Dan Cherry
Ben Crichton
Andrew Davies
David Elmer
Clive Gee
Stephen Green
Angela Luk
Emily Menzies
Chris Naylor

David Paliwoda
Dhanish Patel
Alina Pîrvu
Emily Riggs
Paris Selinas
Olly Simpson
Mark Taylor
Paul Turnock
Jesse Williams

Product Photography:
Dave Branfield
Patrick Quayle

Type set in Frutiger and Fedra Serif.

First published in 2012 in collaboration with
Papadakis Publisher.

An imprint of New Architecture Group Limited
Kimber Studio, Winterbourne, Berkshire,
RG20 8AN, UK

Tel. +44 (0) 1635 248833
info@papadakis.net
www.papadakis.net

ISBN 978-1-906506-32-2

A CIP catalogue of this book is available from the
British Library.

Printed & bound in Great Britain
by Butler Tanner & Dennis Ltd.W

Foreword

Ellen MacArthur DBE

Design is a vision, an answer, a way to move forward - sometimes quite literally. It certainly was the case when Isambard Kingdom Brunel revolutionised the world of transport by imagining an integrated solution, combining trains and steamships, to take passengers from London to New York. This particular journey was fuelled by one man's ideas, and we can only remain in awe of Brunel's technical feats as they symbolise an era of unprecedented progress.

But ideas can also be fuelled by journeys, and my own experience of sailing around the world eventually shifted my thinking, by making me understand the true meaning of the word "finite". Solo racing non-stop around the planet meant that I was constantly aware of my boat's supplies limits, and upon stepping back ashore I began to see that our world was not any different. I started to realise that the linear "Take - Make - Dispose" model that we live by was becoming increasingly maladapted to the reality within which it operates - a reality characterised by materials scarcity and energy prices volatility. We have entered a transition and need to rethink materials, processes, energy generation and even business models.

The circular economy framework, based on two distinct material flows (biological and technical), offers tangible perspectives and a genuine opportunity to reinvent progress. Eradicating the notion of waste presupposes that everything we produce has to fit within the system, whether it is re-entering the technical cycle to be disassembled and re-used, or going back to the soil safely to restore natural capital.

With the advent of consumerism, "design" has too often been considered as a mere marketing tool, a way to increase sales by enhancing products' aesthetics. Yet as the world is standing on the threshold of a new industrial era, that discipline is back at the forefront of potential systemic changes: revolutions in materials science, durable goods designed to be taken apart and be made again, versatile and flexible infrastructure, dwellings that provide services and not only shelter, compostable packaging that feeds agricultural processes...

Brunel shaped his time by thinking big, radical, fiercely new. Our times call for a tidal wave of disruptive innovation, journeys fuelled by design ideas which take the bigger strategic picture into account and lay the foundations of a positive future. Made in Brunel - journeys fuelled by ideas showcases the ideas of the generation of designers and engineers who will be instrumental in taking us there.

Ellen MacArthur

journeys
fuelled by ideas

A founding principle of Made in Brunel was to become an internationally renowned brand representing innovation and excellence; it has grown into a benchmark that we all aspire to reach.

It is an essential part of our motivation, a student led, student fuelled platform for empowering talented people across the whole School of Engineering and Design. Since its inception over 1,000 projects have been launched, all with the support of Brunel University, as well as industrial collaborations and external connections.

"Journeys Fuelled by Ideas" has been the central theme to the 2012 project. It is not just about the destination but the progression of both people and projects and it is exciting to see where these may develop.

For us, this year is nearly over and what a journey it has been, everyone has travelled far and we have endeavoured to give you a taste of our unique pathways in the following pages.

Every year it is amazing to see Made in Brunel go from strength to strength as the work of previous years is built upon. Made in Brunel has always been more than a book; it is a way of thinking. There will continue to be new journeys each year but the same fuel will drive everyone associated with this innovation brand. We are proud to have been a part of Made in Brunel and are excited to become part of its unfolding story.

Ben King and Emily Riggs
2012 Directors

Contents

185 projects

Contents

ecoFrame

Biodegradable photo frames produced from food waste

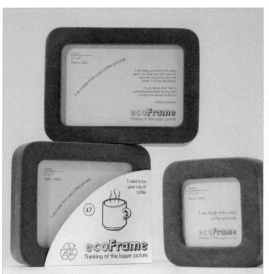

The photo frame market has become somewhat saturated in recent years. Digital frames have captured some market share, so the conventional photo frame is ripe for new development. ecoFrames are manufactured from UK food waste suspended in the biodegradable, starch based polymer, PLA. For the initial product range, used coffee grounds and eggshells were used.

Designed in collaboration with PBFA for Marks & Spencer, the product aims to raise awareness of sustainability within the home. When the user no longer requires the ecoFrame, it can be buried by them where it will fully biodegrade in approximately five years. The protective glass front, card packaging and back are designed for ease of use, sustainability and are fully recyclable.

Hayley Rose Burrows

Product Design

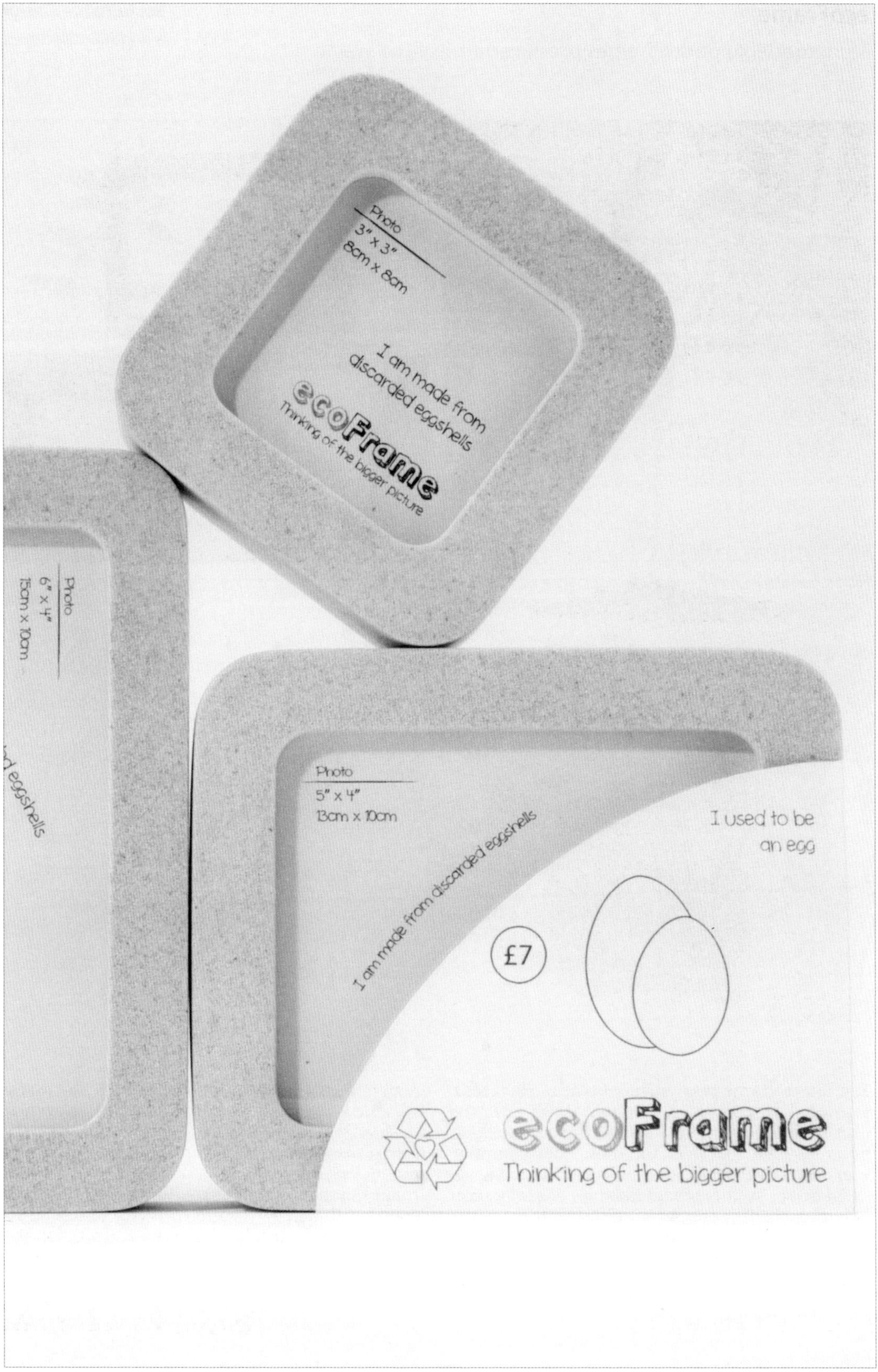
Photo
3" x 3"
8cm x 8cm
I am made from discarded eggshells
ecoFrame
Thinking of the bigger picture
Photo
6" x 4"
15cm x 10cm
Photo
5" x 4"
13cm x 10cm
I am made from discarded eggshells
I used to be an egg
£7
ecoFrame
Thinking of the bigger picture

Pure Plant Pot

A biodegradable seed tray made of starch based bio-plastic

Food waste is always the problem people tend to ignore. Thousands of tons of rotten corn and potato peel are wasted. A number of tea and coffee grounds are thrown away every day after drinking. The starting point of the idea is to reuse those wasted materials and design them back to the soil safely. The idea awakes people's awareness of sustainable lifestyle and influences sustainable behaviour. This is achieved through its hand modelling and planting process, including subconsciously reusing wasted natural materials and creating new life. The pure planting pots provide an indoor planting environment for the protection of seeds and offers an easy approach of planting the entire pot into the ground without getting your hands dirty.

Cheng Yu

Integrated Product Design

Modular Design Ethos - Driving Sustainable Design

Shoe design to act as a benchmark for a modular design ethos

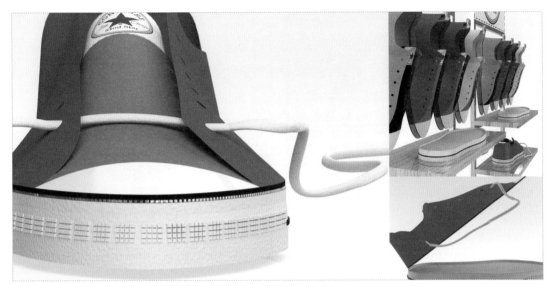

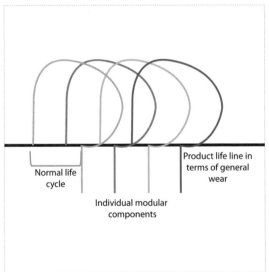

This project is the introduction of a modular design ethos into environments and industries where it would not typically be found. This sets a benchmark that demonstrates how modular design can be used in any industry. The concept shown applies this ethos to the shoe industry where the lower sole area of the shoe is made from long-life polymers which are designed to live through wear and tear; far exceeding the regular life of a shoe. The upper area of the shoe is made from materials geared towards recyclability, which can be interchanged with different styles and fashion driven changes; this ensures the user's continued satisfaction.

Steve Dyson

Integrated Product Design

WasteNot Water Saver

Recycling the water wasted when running the hot tap

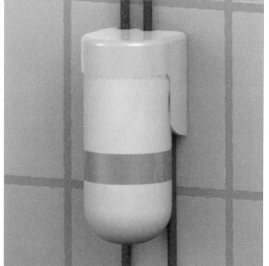

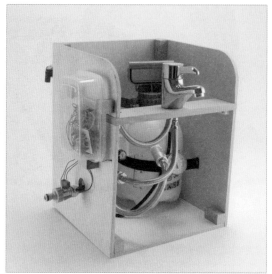

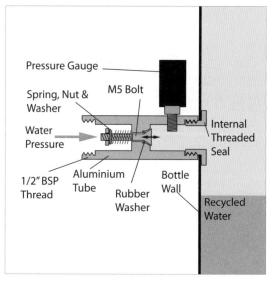

Pressure Gauge

Spring, Nut & Washer

M5 Bolt

Water Pressure

Internal Threaded Seal

1/2" BSP Thread

Aluminium Tube

Rubber Washer

Bottle Wall

Recycled Water

Average UK domestic water use currently stands at 150 litres/person/day of which a large proportion is wasted without being used, e.g. flushing toilets. Overlooked areas of water wastage such as running the hot water taps need to be addressed in order to reach government targets. Cool water needs to be removed from the hot supply pipe before hot water can be accessed.

The WasteNot recycling system separates and stores this water before it exits the tap. The stored water is reintegrated into the cold supply when the tap is opened, requiring only battery power to operate the sensing circuit and diverter valve. Preliminary testing indicates WasteNot could cut average household water usage by 9 litres per day, saving up to 5% off the average water bill.

Mitch Gebbie

Product Design Engineering

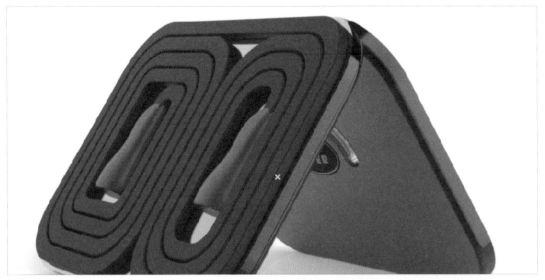

Inspired by a coiled snake and a paper spiral, reCOIL lies in wait in the front passenger's footwell, almost unnoticed except for a folding leather handle and discrete branding embossed on its cover. Like a book it opens revealing an elegant asymmetrical dual spiral cable. reCOIL's form instantly suggests its use, and the cable can be stretched up to 5 metres or left part coiled.

Because this ingenious cable has a rectangular cross-section and its sheath is cast in a spiral form, it has embedded memory and always retracts naturally - aided by magnets and grooves. The user's hands remain clean - unlike the team's hands after a day prototyping the concept with foam and rubber.

S. Chou, C. Conte, Y. Lim, S. Ramm, P. Soltanzadeh, A. Wilson

Design Masters

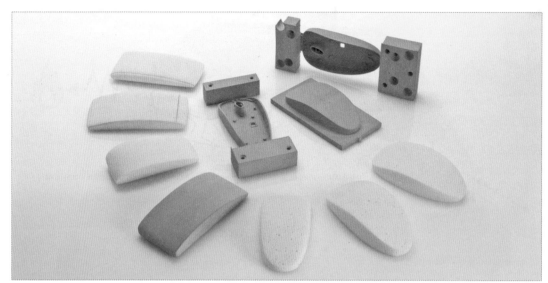

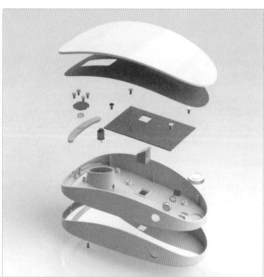

It is estimated that the places where we work, live and grow use 40% of the world's energy and emit 50% of its greenhouse gases (GHG). As considerable progress is being made to reduce GHG emissions in the UK domestic sector, less attention has been given to the commercial sector. In the commercial office environment, office workers do not have a personal interest in energy conservation as they do at home. This project aims to mitigate the trend through a computer peripheral that can be used in the office environment to remind office workers to switch off office equipment. Using haptics to provide tactile feedback and symbolism to communicate reason, Workmate aims to encourage energy conservation among office workers on an individual level and through group-level feedback.

Vincent Ashikordi

Product Design Engineering

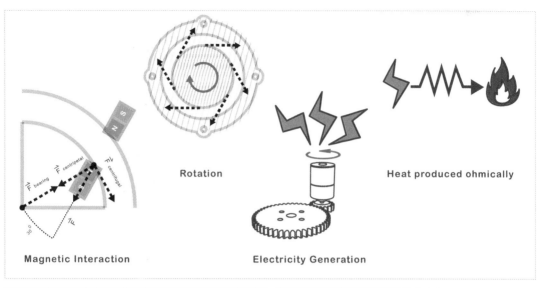

Magnetic Interaction

Rotation

Electricity Generation

Heat produced ohmically

Energy usage is an important issue. Domestic heating was selected as the focus of the Magnetic Heater project, in particular renewable energy, because this is a responsible trend. Elderly people were chosen as the target because according to news reports, cold weather kills thousands of elderly people each year in the UK as they try to economise on their heating bills. Electric heaters in particular are very expensive to run. The basic principles of interaction between magnets have been explored in an effort to develop a new method of energy production. By exploring the feasibility of magnetic motor propulsion, a solution to the complexity of magnetic interaction has been posited by utilising magnetic shielding material to simplify forces involved.

Shu-Yang Chou
Integrated Product Design

Heating Energy Feedback
Changing user behaviour through energy feedback

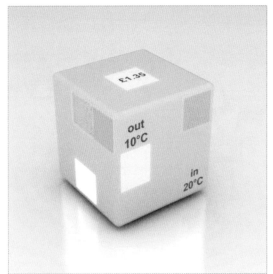

Heating Energy Feedback consists of a group of concepts that provide visual feedback on heating energy consumption, so as to let the user know how much energy they are using. The aim is to produce behaviour change. Four concepts were developed, from different perspectives and for different users. They all give the temperature outside and inside, along with the amount of money spent on heating energy each day. Moreover, each object has an "attention grabbing" system, in order to communicate to the user when they have spent an unusual amount of energy in a day. Finally, they have a data capture system which the user can access to compare the energy spent each month.

Carla Conte
Integrated Product Design

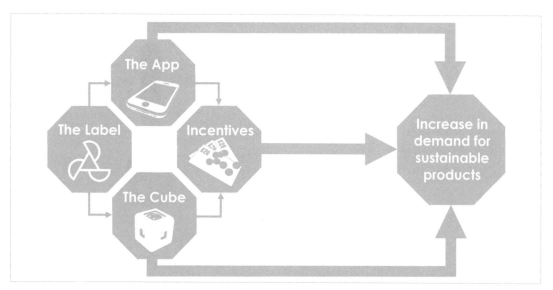

The UK needs to become more sustainable. The Balance brand provides a compulsory label that simply displays whether or not a product is engaging with social and environmental sustainability. The consumer collects Balance points and can show off using the fun and interactive smart phone app or the Balance Cube. Both respond positively or negatively to the 'balance' of the user's product choices and give them a more tangible and immediate overview than the often intangible long term effects of more sustainable choices. The system is complemented by government incentives for good performance. The result is an increase in demand for sustainable product design.

Tim Palmer Fry
Integrated Product Design

Aqua-Q

Rising water consumption awareness through enjoyment experience

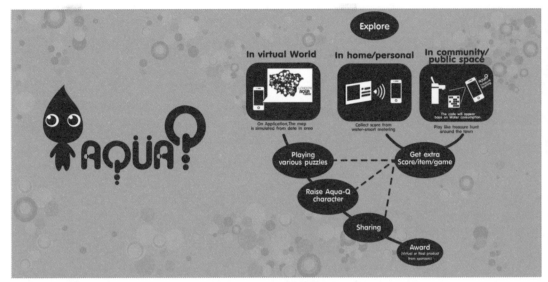

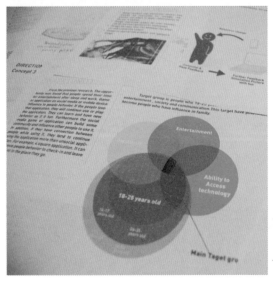

Aqua-Q is a brand and interactive product that encourages users to reduce water consumption through the enjoyment of the experience. It raises user awareness of the problem in the real world through a simulation map with current associated statistics. This system connects the users at home, the virtual world, communities and businesses. In the virtual world application, people can explore and play the puzzle game on the Aqualactic map. At home, Aqua-Q is a water smart-metering system. It converts water usage data into a score. The more water saved, the more points gained. In communities and businesses, Aqua-Q provides measurement for public water usage. People can play, like a treasure hunt, collecting gifts around town.

Sukumal Surichamorn

Integrated Product Design

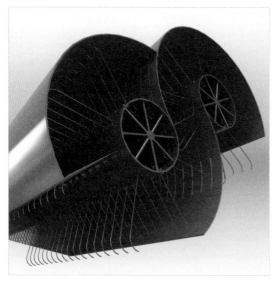

Beach litter has almost doubled over the last 15 years according to the Marine Conservation Society. The demands on the time of local authorities to cleanse their beaches is ever increasing in a throw away society. Councils are working to keep beaches clean by individually picking waste from their beaches as a viable automatic solution for pebbled beaches has not yet been released into the market. The aim of this project has been to provide a faster, easier way to collect litter from pebbled beaches to reduce the time spent cleansing the beaches around the UK and increase its efficiency.

Richard Carter
Industrial Design and Technology

Grapevine

A supermarket 'house wine' bottle refill scheme

START ♡ SHOP ⬚ CHOOSE

RETURN
...helpful bottle holder and SMS reminders

no waste

customer feels special

RETURN

ho**U**se w**I**ne

REFILL

REFILL
...at intuitive machine

UV light and air blade cleansing

up-selling opps while waiting

RINSE ENJOY 🍷 SAVE

RINSE
...quick rinse under the tap at home

bottle reused 30 times

low supermarket staff burden

ENJOY
...a great wine experience

supports small and ethical vineyards

buzz marketing and supermarket brand enhancement

SAVE
...always 10% off equivalent bottle

huge energy and paper savings

customer loyalty increased

Most wine is now drunk at home, making the wine industry look distinctly anachronistic with its heavy disposable bottles, poor branding, fragmented marketplace, questionable transport methods and impenetrable mystique. A new architecture is called for. Grapevine solves these problems by pragmatically placing supermarkets at its core. A choice of six refillable wines from small suppliers changes regularly and is chosen by experts guaranteeing cost saving and great taste. The refill process is quickly embedded into people's weekly routines. Supermarkets benefit from increased loyalty and enhanced brand but the key social benefit is the huge reduction in half-kilo glass bottles being shipped around the world and then needlessly crushed and remoulded.

Simon Ramm

Integrated Product Design

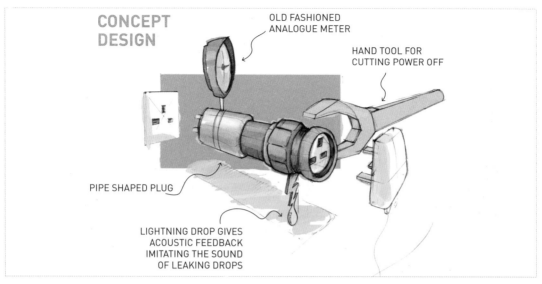

CONCEPT DESIGN

OLD FASHIONED ANALOGUE METER

HAND TOOL FOR CUTTING POWER OFF

PIPE SHAPED PLUG

LIGHTNING DROP GIVES ACOUSTIC FEEDBACK IMITATING THE SOUND OF LEAKING DROPS

Leakage Meter aims to change human behaviour. It balances emotional and behavioural design and attempts to reduce wasted standby power consumption - "electricity leakage". The concept of electricity leakage was used to create a playful, engaging shape. During development the shape was altered several times in an attempt to satisfy safety and behavioural needs: the original pipe-themed concept design gave way to a fire hose and the hand tool was also removed. Leakage Meter informs the user once their device is fully charged using visual and acoustic signals, emanating from the lighting drop. The meter shows the amount of the energy wasted, trying to educate people and foster environmental awareness.

Paris Selinas

Integrated Product Design

Reducing lighting energy consumption through semantics

Hour Glass is a system of switches which replaces mechanical light switches in the home, with no need for any additional changes in the existing wiring. The panel is designed to be placed in a central place in the house, for example the main corridor. The switch has a sensor which is regulated according to season, in order to control the lights efficiently. When you turn the light on, if the system considers the daylight to be sufficient then the hourglass starts and after one hour the light is turned off automatically. In other scenarios the system lets the user know of any wastage. The touch screen central control panel contains a pattern of the house showing graphically in which room the energy is being wasted. It also shows the amount of money which is being spent.

Parisa Soltanzadeh
Integrated Product Design

The aim was to make this project as diverse as possible and promote the aspect of the first 'Sustainable Games'. This illustrates the belief that the only way to protect our planet, is not through the government, or waiting for someone to come along and cure our environmental issues, it is through people acting for themselves. The London 2012 Games give the perfect platform to raise the issue as its groundwork and legacy will attract all the nations around the globe. The character of Patch explores issues of sustainability within this context. Multimedia has not played a vital role in marketing sustainability to date, something this project looks to reduce through investigation and research.

Alkesh Makwana
Multimedia Technology and Design

Pitched Green Roof Tiles

The development of sustainable urban roofing technology

As we place greater emphasis on the well-being of our planet, green roofing has become a force to combat the issues linked with urbanisation. The recreation of natural habitats that would otherwise be lost to construction is part of a larger swing towards a green and sustainable future.

Pitched green roof tile is a product influenced by the current limitations and consumer perceptions of green roofing. Traditional roofing techniques, advanced substrate developments, and purposeful design have allowed this green roofing product to explore an unchartered segment of the market.

James Ward

Industrial Design and Technology

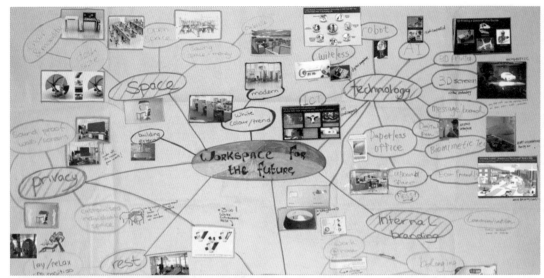

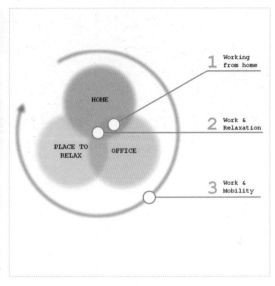

The workspace is the environment where you spend most of your day, yet there have not been many significant changes since the introduction of 20th century office technology. With the availability of next generation technology, the workspace can be transformed into a more productive and creative area. By investigating industry trends, The hope is to further develop the implementation of technology, internal branding, collaboration, health, privacy and the conception of space. The aim is to develop a workspace in which the employees can feel comfortable, more engaged with internal and external colleagues, and above all allowing them to focus. This is intended to increase productivity; benefiting both employers and employees alike.

B. Lee, S. Prasad, M. Shizume, F. Veldhuis, F. Wu, Y. Zalesskaya

Design Masters

Asphalt is a type of composite made of natural aggregates as the structure with bitumen as the binder of these aggregates. This composite has no absorption capability on its own. In order to achieve this capability and property in asphalt concrete, this project reviewed toxic gases that are produced by burning fuels from cars.

As a result, Activated Carbon was added to the structure of Asphalt Concrete due to its extreme porosity and adsorption capability. This process has a small effect per m², but multiplied across the amount of asphalt in use today it could make a significant difference to combating air pollution.

Mohammad Javad Karimianzadeh
Civil Engineering with Sustainability

Construction of a Thames Estuary Airport

Site preparation for a working island platform

London is the economic heart of the United Kingdom and a fulcrum of the global economy. However, this position is under threat with London's airports operating at 99% capacity. This project addresses issues of future air traffic capacity and provides a new international airport hub, sustainably designed to improve London's global transport connectivity with local amenity in mind. The results include a construction scheme for a working artificial island platform upon which airport infrastructure can be built. A holistic project approach has meant that construction methods, environmental impact mitigation and economic growth have all been considered on an equal footing.

B. Bekar, A. Greenland, B. Sidhu, M. Weerakone, V. Yildirim

Civil Engineering with Sustainability

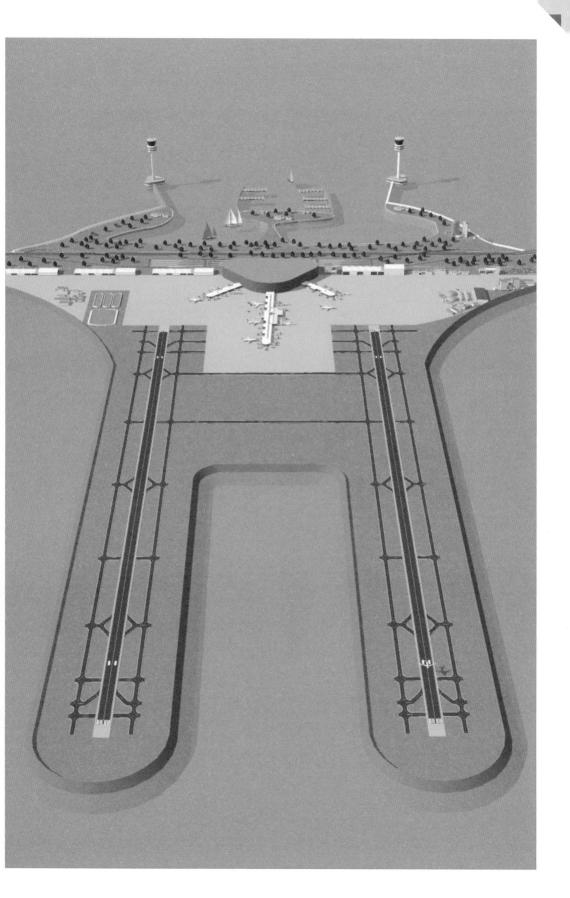

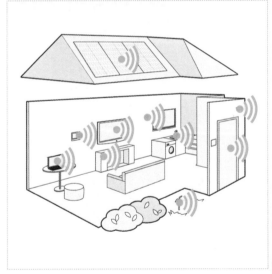

People's lives are getting busier, therefore the home should be a safe place to relax. The Future Smart Home project looks at how technology can be utilised to make home life easier and more comfortable. The Internet, in combination with the growing number of mobile devices - now outnumbering PCs - makes it possible to turn normal homes into smart homes. Today everything is connected and this blurs the limits. Should we be able to control basic home functions remotely, such as heating and lighting? Do we physically need to unlock doors and order groceries, or can this be automated? Where are the boundaries and how is safety and privacy guaranteed throughout this process? Let us use design to find out!

Y. Guo, P. Liarostathi, S. Surichamorn, P. Selinas, O. Tandiroglu, F. Veldhuis

Design Masters

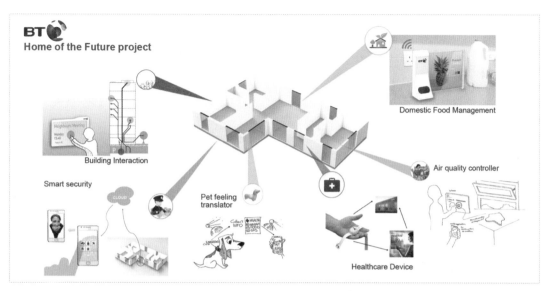

According to recent research by Telefonica there will be more than 50,000 million devices connected to the Internet by 2020. This project, will investigate the changing nature of future technologies and behaviours, especially in the areas of connectivity and networking, exploring the following domains: Daily Life Tasks, Communal Building Networks, Entertainment, Wellbeing/Healthcare, Pet Life, Smart Ecosystem and Sustainability. Conceptual proposals will be used to attempt to project and expand BT's potential in a more dynamic landscape in the near future.

B. Lee, D.J. Lee, Y. Lim, D. Stamatis, Y. Zalesskaya
Design Masters, Industrial Design and Technology

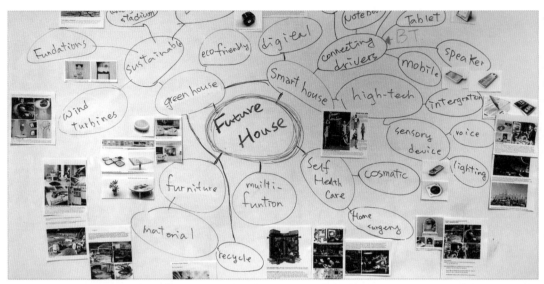

Taking as a starting point the fact that BT has an excellent background in telecommunication services this project intends to suggest future solutions in this area of expertise, developing around the concept of "Home Healthcare". Research on future trends and investigation into potential design gaps have unveiled three design opportunities in the healthcare sector, connecting devices through internet services in order to build an integrated system for better living. The outcome of this project is to connect people with their homes, offering them the security of being taken care of at any time without even noticing.

H.Y. Deng, S.W. Hsu, L. Huang, H.S. Jhang, Y.H. Tien, Y.T. Yeh, E.J. Yu

Overpopulation is the main reason that causes overcrowded urban cities. According to research, there will be 8.8 million people living in London by 2050. The continuous increase in population will lead to serious problems such as lack of space, environmental impacts etc. The number of cities starting to plan for underground space usage is gradually increasing and at the personal level some house owners have started to build large basements under their house. These trends have become the drivers of this project to build an underground hotel at the centre of London, providing a convenient location, affordable prices and connections to other underground facilities.

S. Baghchesarai, M. Li, X.M. Liu, S.Y Mo, W.S.M. Poon, Y. J. Wang
Design Masters

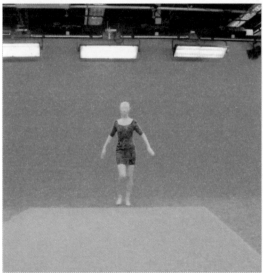

Shop window visual merchandising exists today as a result of department stores opening and creating fierce competition such as 'Le Bon Marché', the world's first department store dating back to 1852. However, the techniques used have barely changed in over 150 years, with mannequins in rectangular surroundings still visible on every high street. Commercial Urban Screens looks to revolutionize visual merchandising by pushing current restrained digital screen uses further by using full scale projections. By using a short throw projector and a window film to hold the image, physical construction constraints are eradicated, making any display possible for this form of street art.

Craig Aburrow
Broadcast Media Design and Technology

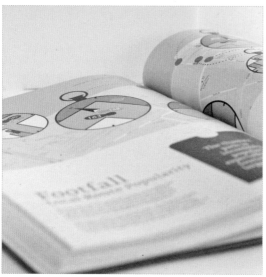

Improving a venue's wayfinding is a clear and effective way of improving the visitors' overall event or exhibition experience. In this project, many tools were implemented such as footfall analysis, visibility splays and visitor-specific flow routes. The tools were developed to try and build a picture of the average visitor, defining key aspects such as where they will be traveling, by what method, and their desired destinations, around and in the event. The final documents were designed to contain transferrable information so future hosts could also use the pack to realise how wayfinding can be improved around their event.

T.Logg, D.Patel, J.Topp, P. Verheul
Industrial Design and Technology

Designer Babies

Is it one step too far?

Since 1953 when Watson and Crick discovered the structure of DNA, the use of technology and design to monitor and manipulate genetics has continued to grow and develop. A Wellcome Trust Poll in 1997 reported that 37% of people thought that human genetics research was already going too far, so what about the other 63%?

Services are already available to enable the creation of designer babies with a range of clinics offering Primplantation Genetic Diagnosis (PGD) and In Vitro Fertilization (IVF) to screen embryos for life altering or threatening diseases before the embryo is implanted into the mother.

> *"We share 98% of genetic material with chimpanzees and 51% with yeast - genes do not tell us the meaning of life"*
>
> Tom Shakespeare

Should the line be drawn at the prevention of babies being born with potentially life threatening diseases that not only affect them but also the families they are born into? Or with the selection of eye and hair colour as has previously (although no longer) been offered by the Fertility Institutes? Or is it when more options such as height, weight and even personality are available for customisation? Although currently technology is only available to enable the parents to choose traits which one of them possesses - for example they could only select a blue eyed baby to be born from an embryo created by them if one of them had blue eyes (Kleiner, 2009) it is unknown how technology and knowledge of human genetics will continue to develop, enabling more possibilities.

Do we already create 'designer babies' to a certain extent purely due to who we choose to mate with? People tend to be attracted to people with characteristics they would wish their children to possess and in mating with these people they are increasing the probability of their children possessing the desired traits.

The Fertility Institutes in America already offer sex selection with a 99.9% accuracy and when asked about the demand a spokesperson remarked: "The demand is very big. We have seen couples from every country on earth and also a scientific research couple from the North Pole!" The Fertility Institutes, New York.

Genetic engineering can allow parents to have the 'child they have always wanted' and there are many examples of mothers having a fourth child in the desperate hope to have the longed for girl or boy. I am sure that they would have loved the chance to know the sex of an embryo before it was implanted. Unfortunately for them, sex selection is only currently legal in the UK for medical reasons, for example if there is a genetic disease that only a male can be born with. The World Health Organisation state that there are three main reasons for sex selection - medical, family balancing and gender preference (WHO, 2012). However does it really make a difference to the child if they are born a girl or boy? Is it selfish that the parents can choose how they would like their children to be born?

There are a number of natural methods available online which allegedly improve chances of delivering a baby of the desired sex. Even Aristole was reported to believe that if a man tied his left testicle before intercourse, a male would be conceived (Dr Kaplan, 2009). There is also a very serious element to gender selection, making it legal could help to prevent exiting, illegal sex selection methods and cases of sex selective abortions being carried out. The Global Change report in 2011 stated that in India and China a number of newborn babies are murdered because they are the 'wrong gender' and it was reported recently that:

"Some 50,000 female foetuses are aborted every month in India. Baby girls are often killed at birth, either thrown into rivers, or left to die in garbage dumps. Its estimated that one million girls in India "disappear" every year."

(Vargas, 2011) Would it be better to ensure that the parents were implanted with an embryo of the desired sex rather than this occurring?

PGD can help to screen embryos to see if a child has the genes for an inherited condition and give the parents the choice of whether to go ahead with implanting the embryo. There are many examples of this, with PGD being used for preventing diseases that could have been life threatening or in some cases even deadly to the child born. Even if it does seem to be 'playing with nature' wouldn't we all prefer to provide the safest possible options for our children so that they could go through life with less problems?

In England and Wales 1 in 5 pregnancies end in abortion. If parents are having IVF, then embryos without any genetic abnormalities could be selected. If the health of a child is likely to be significantly poor, or require a large amount of complex care, should be able to foresee such a situation? Are characteristics such as hair and eye colour as important as health; is having blonde hair and blue eyes as important as having perfect sight or perfect hearing? Agnes Fletcher remarked how many deaf people would not regard themselves as being disabled, instead they consider themselves to be a minority (Lee et al., 2002) We are on the brink of technology offering parents the option to select their personal preferences. In some cases the genetic selection of a child can not only prevent them from being born with any inherited conditions but also benefit their living sibling. The new baby can be born as a 'saviour sibling' to become a tissue donor to their living brother or sister, enabling them to prolong and potentially save their life. Adam Nash is a prime example of this, a test tube baby born with matching tissue to his older sister so that he could save her life through bone-marrow transplants. These were necessary as she had a dangerous blood disorder.

It brings us back to the Nature/Nurture debate. Some characteristics such as eye or hair colour will be easier to predict and offer as a choice with more certainty. Others such as being able to choose an embryo which may be more athletic or intelligent could be more difficult. Just because someone has been born to have a certain trait due to gene selection there are a number of other factors that could affect it. If you selected an embryo that should in theory be more intelligent and then when the clever child was born never educated them, their intelligence is unlikely to develop. This can be seen with identical twins. Even if they are born identical genetically, after a short amount of time differences in their behaviour and personalities start to appear.

Will what is deemed desirable at the time of conception still be the same when the child is older with fashion and trends frequently changing. Tattooing, intentional scarring and piercing are presently very popular ways to modify our own bodies. Some may think of genetic engineering as the 'Ultimate Tattoo' are we really entitled and informed enough to make such life changing decisions for our offspring? However if the decision was in the case of height, for example with Loran Dwarfs, would it cause them less problems in later life if a gene was discovered to modify height and design them to be 5 feet 4 inches instead of 3 feet and 2 inches?

Where will it go next, might we be born with two livers in case we drink too much and destroy one as the 4,160 who died due to alcoholic liver disease in 2005 did. If the UK does not legalise more genetic engineering, will people keep going abroad or receive treatments in illegal practices? We need to have a very clear view of where this is leading; be very careful about where restrictions should be put in place without constraining any possible medical advances that could enhance life.

The Fertility Institutes when asked how far is too far remarked: "We hope no one ever gives us a list of what we can and cannot offer."

If you knew that your child would have a 40-50% chance of developing an inheritable disease that would ultimately cause them great harm, would you want someone, somewhere, to have informed genetic research groups to not investigate a genetic cure?

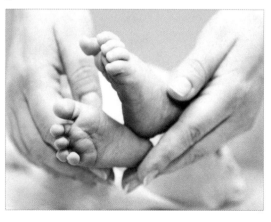

Emily Riggs
Product Design

Infant Oral Motor Development Toy

A germ-free product for babies to enjoy and experience

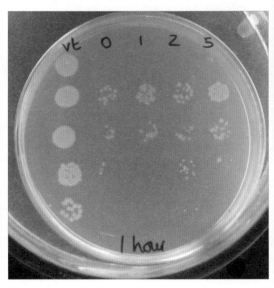
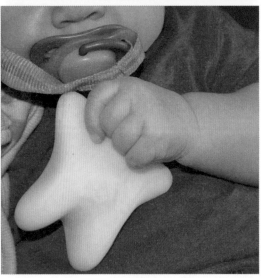
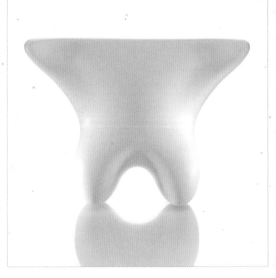

A baby's mouth is often referred to as their window to the world. This is true in terms of oral sensory development and also of a baby's vulnerability to germs and bacteria. The current oral baby product market is highly saturated with many variations of essentially the same form; however, there are none that promote antimicrobial properties and their benefits. This project involved developing a deep understanding of the injection moulding process, along with the antibacterial surface testing of an enhanced material, in order to explore potential applications of this material for the development of an improved oral baby product.

Laura Ginn

Product Design Engineering

infant nose
depth...
0-3-5 months.
14.6/4 = 3.65cm.
46-19 months.
16.7/3 = 5.56cm.

Length
or legs

Goldilocks

Projection mapping film

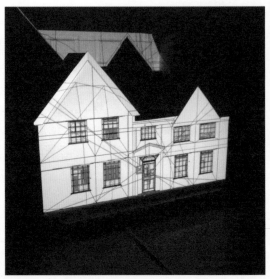

Goldilocks is a proof of concept short film, incorporating both live action and 3D animation, designed to be projected onto the exterior of a real life location. The action takes place inside a 3D replica of a real building; the premise involving a baby sitter telling the story of Goldilocks to a child with the story coming to life around them.

Whilst 3D projection mapping techniques are regularly used for publicity stunts, there have been very few attempts to translate a conventional story into a large scale projection experience, and it is the possibilities that lie within this area that are explored with this film.

Jordan Fisher

Multimedia Technology and Design

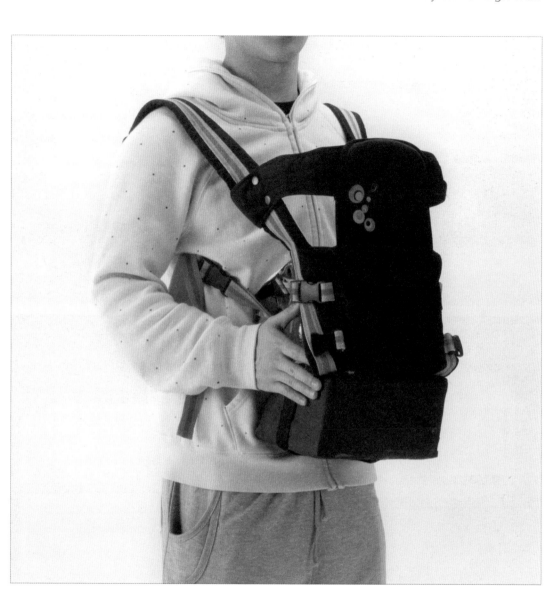

A baby carrier is a convenient way of carrying a young child, which can give the baby and parents the comfort of being close together, while leaving the parent's hands free to get on with other things. A baby is born with the need to be loved and never outgrows it, but problems can appear when the parents need to take care of the baby and complete other tasks at the same time. This project is aimed at designing a new type of baby carrier, which contain muti-functional and improved protection through design that will give the users easy, comfortable and safe journeys.

Benny Ho Hon Chi
Industrial Design and Technology

Playhaler

A stress reducing MDI asthma administration device for young children

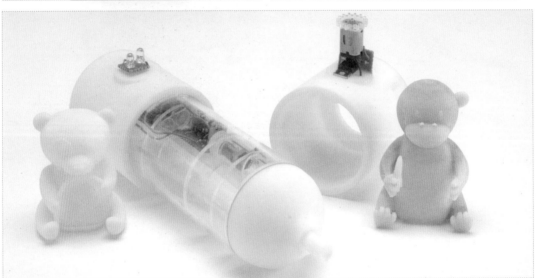

In the UK 1.1 million children suffer with asthma. This shocking figure shows a momentous need for reliable and efficient means of controlling asthma in children. The Playhaler is aimed at reducing stress during administration of MDI aerosol medication to children 1-3 years. The device is extremely important in controlling asthma in young children and substantial research has revealed that existing devices are not controlling the condition effectively. Playhaler tackles child behavioural issues and parent/caregiver administration issues. Playhaler encourages children to accept the medication by incorporating a playful, enjoyable process, providing parents with an intuitive, easily transferable device.

Matthew Bastow

Industrial Design and Technology

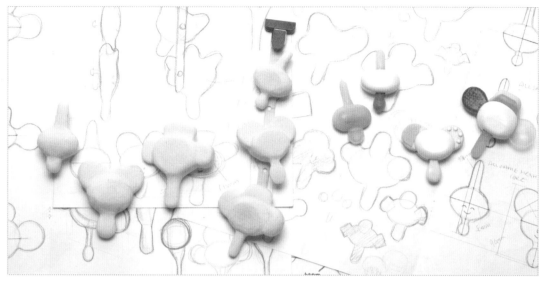

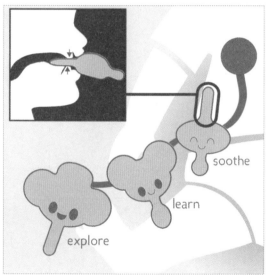

In collaboration with the Speech Therapy team at Guy's and St Thomas' Hospital, the project aims to promote the development of a healthy mouth from birth. Despite a highly saturated teething toy market, parents remain relatively unaware of the important role of the mouth in infant development. Teethers are designed to reduce teething discomfort, whereas 'nom noms' are designed to support the whole mouth through the growing stages of oral motor and sensory development. The 'nom noms' family, used in turn, help the infant 'Soothe', 'Learn' and 'Explore', promoting healthy development and confidence in feeding and chewing skills.

Sophie O'Kelly
Product Design

The Diversity of Alumni

Made in Brunel is a globally recognised brand, with a history of international partnerships from Asia to the Americas. Our alumni not only represent Brunel as an internationally-acclaimed university, but also work to forge strong links with industry around the world. Brunel alumni are truly global thinkers.

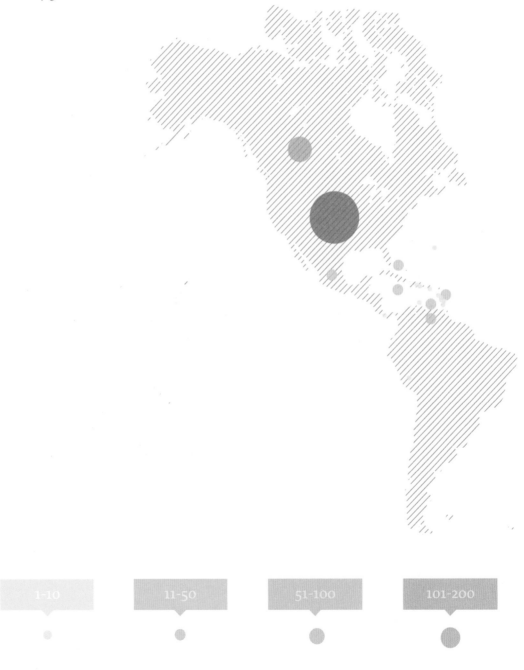

| 1-10 | 11-50 | 51-100 | 101-200 |

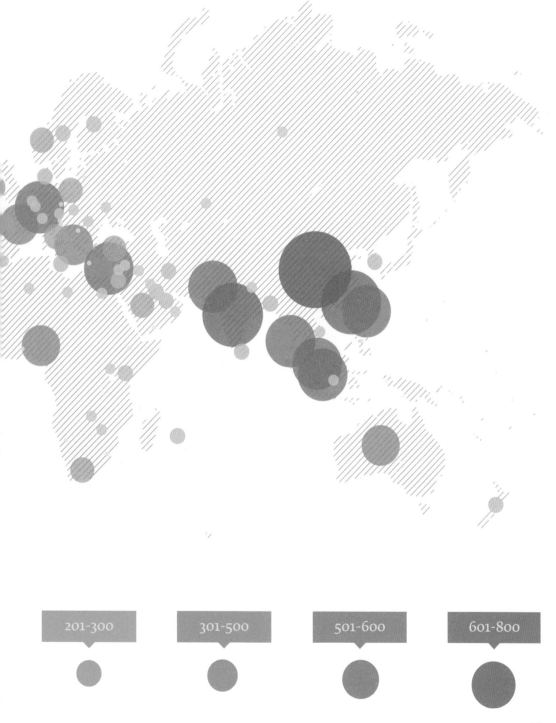

201-300   301-500   501-600   601-800

Groponics
Children's hydroponic gardening system

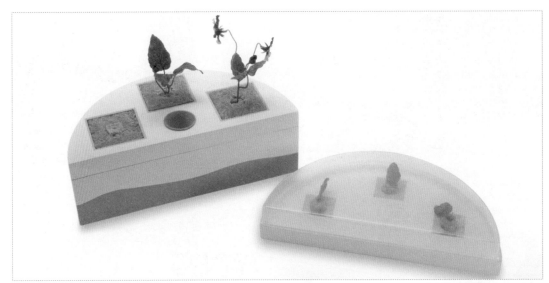

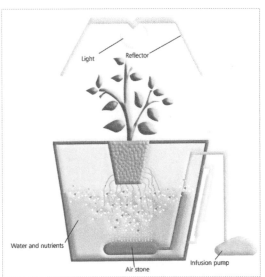

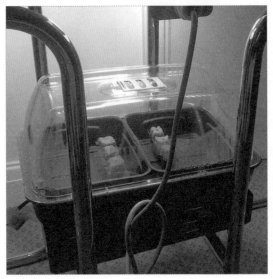

With children spending less time in natural spaces than their parents did when they were younger, there is a danger of children having less appreciation for the natural world due to their lack of first hand experience of these areas. Groponics is a hydroponic gardening system for children aged 8-11. Its aim is to improve children's appreciation and knowledge of the natural world using a fun, hands on, approach, allowing children to grow a variety of plants themselves. Groponics is a learning toy with supporting features including a website that provides engaging content, allowing children to broaden their knowledge of nature.

Jaymini Desai
Industrial Design and Technology

Educational Video for LIFESAVER Systems
Teaching children about the lack of safe drinking water worldwide

This project is an educational video made for LIFESAVER systems. It will be used within schools to help effectively communicate the issue of the shortage of safe drinking water across the world and explain the solution of LIFESAVER technology. It has been designed to be fast-paced to keep a young audience engaged whilst using a variety of animation techniques to tell the story. For parts of the edit the visual style is fun and simplistic with crisp vector imagery, whilst other sections are hand drawn using rotoscoping techniques, with the simplistic style complimenting the poignant message.

Rachael Greene
Broadcast Media Design and Technology

Flip Books Project

Gamification to improve young people's learning within the UK

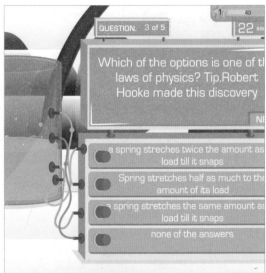
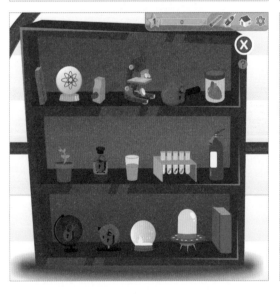

This project aims to aid schools' approach to learning, through strategies, conventions and implementation of emerging technology found within the continuously evolving gaming industry; moving away from conventional methods with their limited interaction. The "Flip Books Project" allows pupils to enter a social gaming environment and complete challenges to help accomplish an overall mission within subjects, in an interactive, rewarding and challenging space. The first example focuses on the subject area "Science" to simulate the functionality. To stimulate active learning, a physical book is supplied, which will display virtual content using Augmented Reality technology making learning simpler, more enjoyable and worthwhile.

Farha Rehnnuma
Multimedia Technology and Design

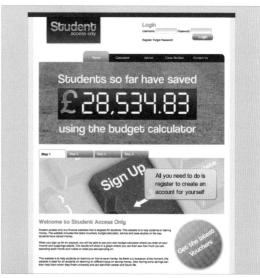

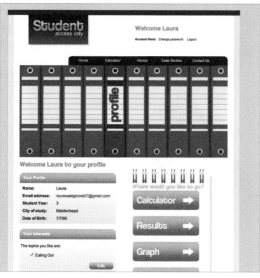

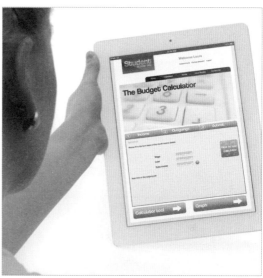

This project is a student finance portal aiming to tackle the growing problem of student debt, often reaching as high as £50,000 post-graduation. This website will help students on saving money through tools and advice on ways to save. Current methods by which students save money and seek guidance online are not successful. This site features a budget calculator which will allow students to login to their account each month and add their total income and outgoings. Each month student users see a progress chart on how much money they have saved and what to do next. In addition, the website will display different voucher codes that interest the user that can be used online or in stores.

Laura Castiglione
Multimedia Technology and Design

Fish Eye - The Making Community for Film Makers

A study in open innovation and collaborative design

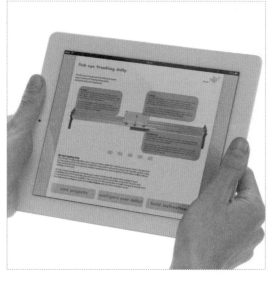

User innovators have an abundance of new technology to help them 'make' things. But why have not we seen an explosion of fully developed products developed by them? It is not just the physical tools that are required when developing new products. Critical design thinking and an objective design process must not be undervalued. By working with groups of user innovators in two collaborative case studies, a community website, Fish Eye, was formed. Based around a camera tracking dolly it provides a platform to collaborate together and design new products that meet the specific needs of those who develop them. As a result of the community based design model, users will learn new physical skills and improved critical design thinking.

Edwin Foote

Product Design Engineering

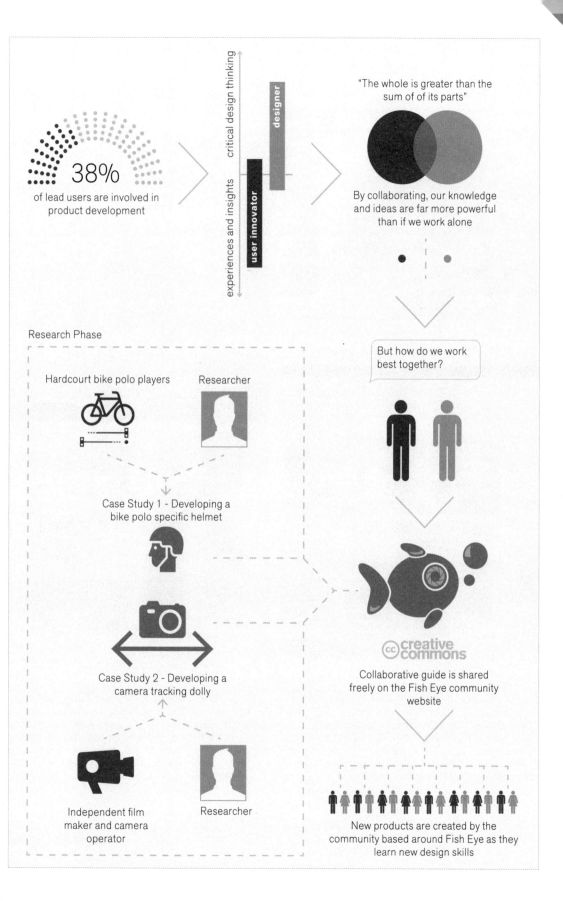

38%
of lead users are involved in product development

critical design thinking

designer

user innovator

experiences and insights

"The whole is greater than the sum of of its parts"

By collaborating, our knowledge and ideas are far more powerful than if we work alone

But how do we work best together?

Research Phase

Hardcourt bike polo players   Researcher

Case Study 1 - Developing a bike polo specific helmet

Case Study 2 - Developing a camera tracking dolly

Independent film maker and camera operator   Researcher

(cc) creative commons

Collaborative guide is shared freely on the Fish Eye community website

New products are created by the community based around Fish Eye as they learn new design skills

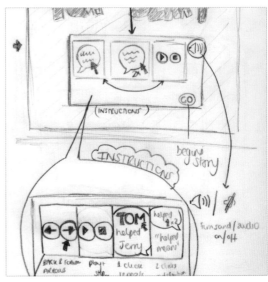

InteRead is an online-based interactive tool for primary schools. The site aims to enhance the interest of the En2 reading curriculum with the use of popular licensed characters. Working closely with Warner Bros, we can see the influence these properties have on children and the possibilities this could have when merging them with e-learning. The site consists of interactive comics based around the stories of licensed characters, which allow pupils to develop their reading skills and confidence. Assisting them are also games, activities and a 'teacher's corner', supplying teachers with lesson plans, worksheets and ways to interact with the site.

Beth Williams
Multimedia Technology and Design

Starting at university can be a daunting milestone and there are many generic freshers week survival guides available, but few universities have their own specific ones. Fewer still have guides catering specifically for the needs of disabled students. A study of Brunel from a wheelchair user's point of view was conducted and from this research a pocket sized graphical guide to Brunel was created. It contains lots of general tips about starting out at Brunel and also focuses on the particular areas and aspects of the University that would be the most important during the partying of freshers week, such as visual markers to locate the ramps in the crowded union clubs and bars while slightly worse for wear!

Joe Carling
Industrial Design and Technology

Making a Mess
Educational reform and the role of creativity

Sir Ken Robinson, an internationally recognised leader in the development of education, creativity and innovation, puts forward the case that during our time at school we are "steered benignly away from subjects, things we liked at school, on the grounds that you would never get a job doing that". I can't help wonder if today's approaches to learning ensures that children's' creativity is encouraged and whether they are able to pursue what they love, if at all.

Today the world is facing unprecedented challenges; multifaceted economic/financial, social, climatic, and environmental crises - all needing creative and innovative solutions. We need to have the capacity to think divergently and 'out of the box' for such complex challenges. The world is demanding the very skills that our educational system at present is unintentionally stifling within children.

> "When students find something they enjoy and can excel in, they do better in education generally"
>
> Ken Robinson

Stimulating a child with creative endeavours, captivating their imagination and getting them excited about what they are learning, not only equips them with necessary life skills, but it can have a catalytic effect that heightens their attitude towards school overall. School becomes a place where they want to be, where they can't wait to learn. Surely creative environments also have an effect on behaviour, encouraging them to be sensitive to world issues and those around them, improve disruptive behaviour because they are less bored and develop their productivity.

A case study of Finland's approaches to learning highlights that shirking constant exams and rigid approaches to teaching, while encompassing creativity, can do wonders for children's learning. Finnish students score near the very top on international tests, yet they do not follow the Asian model of study, study and more study. Instead they start school a year later than in most countries, emphasize creative work and shun tests for most of the year. The study also notes that Finnish teachers are paid well and treated with the same professional respect as doctors or lawyers. It seems that nurturing creativity and optimising the quality of our teachers are vital in order to raise educational standards.

English children are now the most tested children in the industrialised world; with pupils being subjected to over 70 tests during their school career. But it's not only students who have felt the burden of the increased number of examinations. For schools, the costs to sit exams have dramatically risen with spending on exam fees nearly doubled, to over £300m, between 2002 and 2010. For teachers, many are forced to 'teach to the test' which can be deeply disheartening and un-motivating; resulting in teaching becoming dull, narrow and uninspiring. There is no incentive or reward for creativity, only results. And for pupils, tests are damaging their motivation and enjoyment at school in general, whilst they are also a well-documented source of stress and anxiety.

It is important to note here, that exams themselves do not contribute to the raise in standards of learning. Some are of the view that they are purely to measure how much information a child has been able to retain, with no inclination of the qualities indicative of a well rounded education - from independent reasoning to creative thinking. It is questionable how well testing actually assesses whether a child has mastered the content they are being examined upon or not. In the words of the US educational psychologist Joseph Renzulli, we have created a new version of the Three Rs: "ram, remember, regurgitate".

Sir James Dyson states that "great teachers are the single most important factor in successful teaching" and this is the key to unlocking a wealth of talent, curiosity and passionate desire to learn. A great teacher is one who not only has great knowledge of their field, but one who also takes risks when engaging young minds and sparking enthusiasm for the subject. Teachers should be given the freedom to use their imagination, to tailor lessons around what they think will be the most effective ways of communicating the curriculum. Robert J Sternberg suggests that "The most powerful way to develop creativity in your students is to be a role model. Children develop creativity not when you tell them to, but when you show them."

Guardian reporter, Adam Lopez, delves deeper into the idea of teaching in his article 'What makes a brilliant teacher?' by pinpointing the sense of awe and wonder, great teachers create to develop enquiring minds "with an insatiable thirst for learning that endures. This is ultimately the teacher's ability to progressively build and reinforce high quality educational attachment relationships, which in its infancy can be termed rapport." This allows them a "way into the pupils' worlds" and become a pivotal milestone to help boost a child's confidence, develop their appetite for a subject and encourage them to think creatively and freely. Teachers that perform to high levels such as this are a valuable commodity to our society, and have the opportunity to help shape future generations with the right training and support from Head Teachers and government policy makers. After all;

> *"Students placed with high performing teachers consistently progress three times faster than those placed with low performing teachers"*
>
> *James Dyson*

One of the most recent changes to the educational system was made recently when the Ofsted Inspection Framework was implemented after being published in March 2011. Regarding the 'The Quality of Teaching' Ofsted propose to judge teaching standards based on a set list of criteria. One of which is how well the teachers "enthuse, engage and motivate pupils so that they learn and make progress." This will hopefully allow teachers some wiggle room when it comes to the methods they use to engage and motivate students, whilst also being able to create challenges for them to fuel their curiosity further, all of which is imperative to the inclusion of creativity within the classroom. Some schools are starting to adopt philosophies themselves that are changing the misconceptions of creativity. One case study is that of Christchurch primary school in Essex, whereby "creativity isn't shoe-horned into the curriculum - it is a part of daily life" insists Head Teacher Kevin Baskill. Just one example of this are the year 6 pupils taking a slightly different approach to learning about numeracy and literacy by making cakes. They write out the instructions, convert the measurements, add the weights and end up with lovely cupcakes. Mr Baskill adds "It's those sorts of creative things that you remember, not the endless, dry, tedious revision papers."

Evidence of creative learning can also be found in Wroxham School in Hertfordshire. Head teacher Alison Peacock transformed the school from being in special measures to outstanding in three years. Ms Peacock says a large part of the success is down to involving as many external artists, actors, musicians and poets in the fabric of the school as possible; "I could have invested in literary consultants, but I wanted to show the staff the art of the possible - that we could create an entirely new environment by learning from artists and from each other"

There are many arguments for embedding creativity into the curriculum. Some are already reaching schools and teachers; bold efforts and successes to inspire other institutions to follow in the same footsteps. The coalition government is striving to make headway on reforming the education system as we know it, with various plans already being implemented where creativity in the classrooms is to be encouraged. 2012/13 will unquestionably be critical years for educating Britain's children. It is as Sir Ken Robinson so intriguingly puts it: "Transforming education is not easy but the price of failure is more than we can afford, while the benefits of success are more than we can imagine."

Sarah Hutley

Industrial Design and Technology

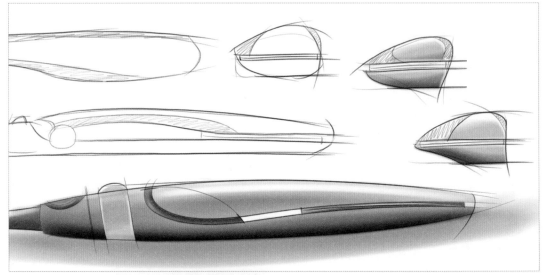

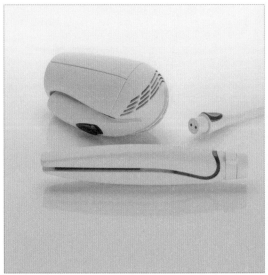

The project, in collaboration with PBFA, was to design and develop a hair dryer and hair straighteners for a highly compact and portable travel set to be sold in Marks & Spencer. It features a removable, interchangeable cable for use on both products. The products offer space saving advantages over current products, which often neglect the need for an effective cable management solution. The rotary switch on the hair straighteners is another USP which offers advantages of convenience and reassurance. They have been styled as a range, and draw influence from a more delicate and refined approach to design detail.

Olly Simpson
Product Design Engineering

Paiya

Wheel water filtration and hand washing device for rural India

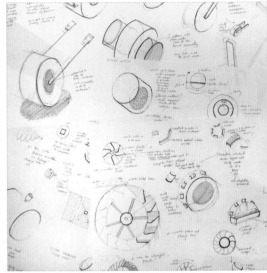

Every twenty seconds a child dies from water borne illness. Travelling in rural India revealed the unsafe nature of the water supplies and children's lack of personal hygiene. Research indicated that hygiene-related diseases can be prevented by small, but significant, changes such as washing hands regularly and drinking clean water. This influenced the design of a water filtration and hand washing device directed at children. By rolling the product along the ground instead of carrying it, the form of its wheel filters the water contained inside in motion to provide an enjoyable and educational experience, making the task of collecting water easier.

Jyoti Lakshimi Sharma

Product Design Engineering

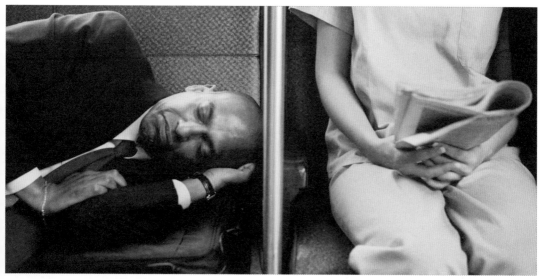

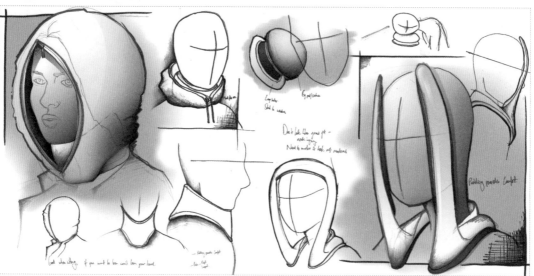

Modern day hectic lifestyles mean that, in the UK, 75% of the population wake up exhausted, endangering their health and leading to a lack of motivation. This affects relationships, work and many other aspects of people's everyday lives. This product aims to combat this by helping people 'power nap' whenever and wherever they may be, utilising human-centred design and deeper understanding of sleep. This is achieved by effectively supporting the head and neck in an ergonomically comfortable position, helping the user to relax and comfortably fall asleep while in an upright seated position.

Luke Gray

Industrial Design and Technology

Launch

Customising the way that you wake up every morning

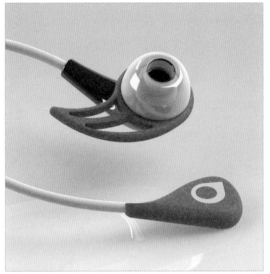

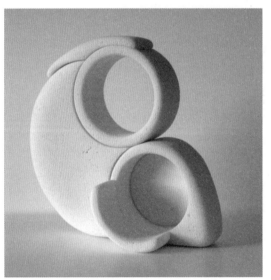
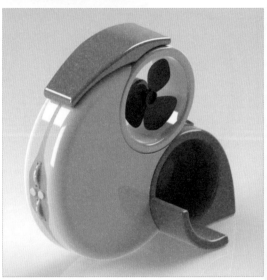

With working lifestyles becoming more hectic, people are finding it more challenging to wake up on time. There are a variety of options available for waking up, but not everything works for everyone. Launch aims to combat the problems associated with waking up abruptly, by allowing each of us to customise our waking up experience to our needs through a range of different stimuli, including light, sound, vibration and air. This allows people to wake up more easily and refreshed, reducing the disturbance caused to others around them through a lightweight earpiece worn discreetly as they sleep. With Launch, everyone can experiment with how they wake up. Launch - how do you want to wake up?

Jamie Topp

Industrial Design and Technology

HOW
DO YOU
WANT TO
· WAKE UP?

Like   15k   Send   Tweet   4,080   +1   4

Moved to using the light this morning - what a difference!	Has anyone tried the fan and light together? Works a treat, so relaxed.	Long night last night, got up early thanks to Launch.	Nothing better than waking up to a bit of Queen, We are the Champions.
Jim, UK - 32	Matt, UK - 25	Thomas, UK - 21	Sarah, UK - 29

LAUNCH OUT OF BED EVERYDAY

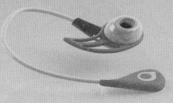

The Element	Base Station	Wireless Charging
The first wireless earphone you can wear to sleep, and it can wake you up via your favourite music, as well as giving a slight tickle to wake you up out of that deep sleep.	If the ultra-bright white light is not enough, the gentle breeze from the inbuilt fan will wake you up feeling refreshed as though you have just woken up on the beach.	Get up and out of bed, no need to find the socket or get the right cable. Just place the element into its charging bay, and it will start charging automatically.

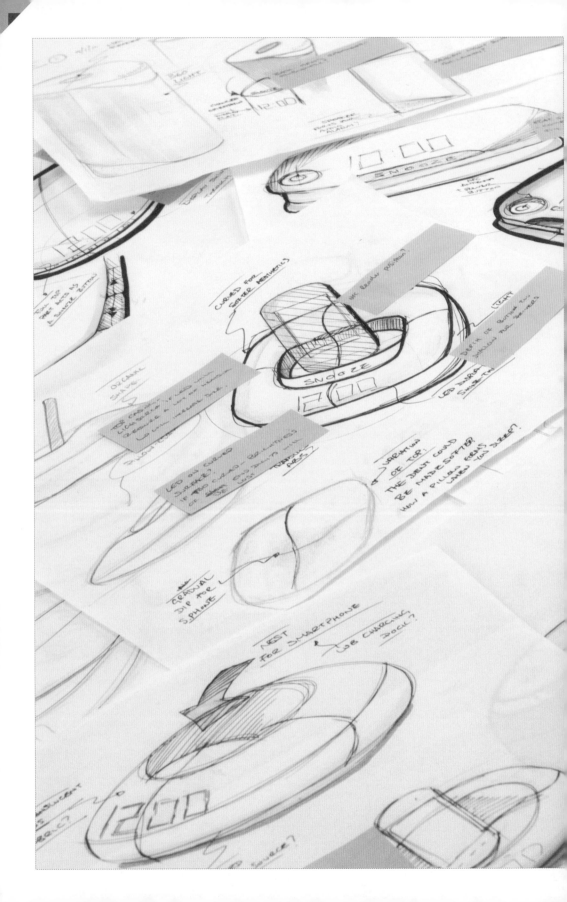

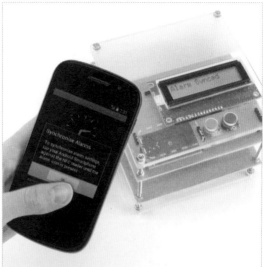

Over 70% of Mobile phone/Smartphone owners use their devices for morning alarm purposes, instead of digital alarm clocks. To respond to this trend, a Smartphone application and Near Field Communication (NFC) technology were integrated, by allowing wireless synchronisation of alarm data, in order to match current user behaviour with the aim of providing a more intuitive interface. The secondary aspect of the product aims to improve user awakening as current audio alarms result in low alertness and poor user experience. Based on research showing light's effectiveness as a stimulus, the product utilises a specific wavelength as a wake up method to improve the well-being and alertness of the person each morning.

Angela Luk
Product Design Engineering

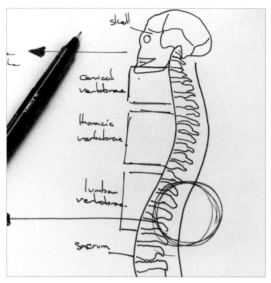

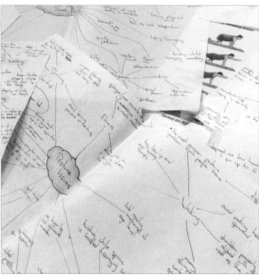

Childhood back pain is currently at an all-time high with 48.5% of secondary school pupils in the UK already having some form of back pain. To address the challenge of children and students being required to carry increasingly heavy loads on a daily basis, this project has developed as a redesign of one of the most popular backpack styles currently on the market. The redesign focused on distributing the weight around the back more effectively in order to reduce the amount of stress experienced by the wearer and minimise the chances of developing back pain.

John Aston

Industrial Design and Technology

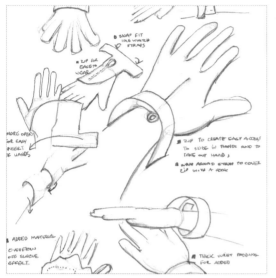
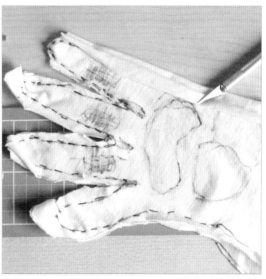

By 2034, one in two people are going to be aged over 50. The ageing population needs gloves which not only protect but enhance performance for the ageing hands. This project aims to improve the daily lives of the elderly who are on the move in a mobile society. The glove is based on research and analysis of products being used whilst outdoors and on the go. Eliminating today's frustrations and steering away from stigma through good design is important, to bring desirability to make the product attractive to the users.

Nejal Patel
Industrial Design and Technology

Placement Diversity

We know that one of the best ways of learning is by doing, which is why the Brunel team and us invest a lot of time and effort into finding the best placements. Our achievements have not gone unnoticed, as we have been awarded the UK's best Placement and Careers Service at the National Placement and Internship Awards.

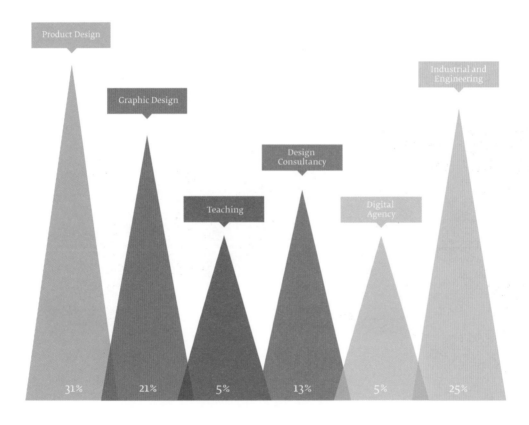

Product Design — 31%
Graphic Design — 21%
Teaching — 5%
Design Consultancy — 13%
Digital Agency — 5%
Industrial and Engineering — 25%

65 placements in 2011

80 placements in 2012

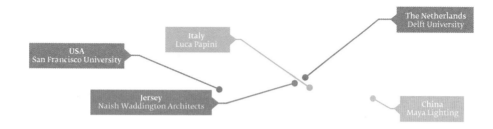

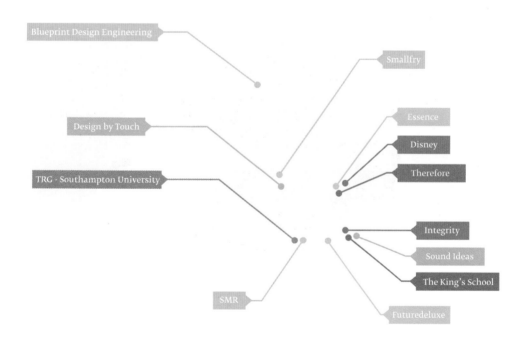

The Netherlands
Delft University

Italy
Luca Papini

USA
San Francisco University

Jersey
Naish Waddington Architects

China
Maya Lighting

Blueprint Design Engineering

Smallfry

Design by Touch

Essence

Disney

Therefore

TRG - Southampton University

Integrity

Sound Ideas

The King's School

SMR

Futuredeluxe

90+ placements expected in 2013

The Lunch Purse
An elegant women's lunch bag and container system

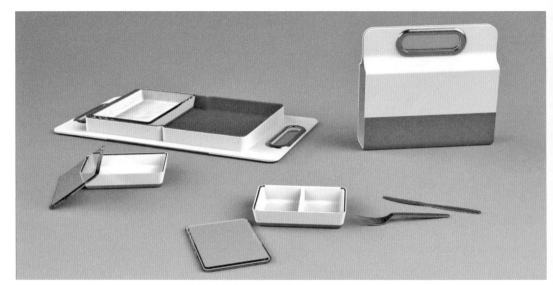

Increasingly women are taking their own lunch into work in a bid to save money and control exactly what goes into the food they eat. However current lunch bags on the market are poorly tailored to their needs and fashion taste, leading many to believe they can do without one. The Lunch Purse provides a sophisticated alternative that integrates with their busy lifestyles and caters to a more discriminating sense of style. The slim bag unfolds to unveil its food containers within, creating a structured space for users to enjoy their food, whether it's at their desk or on the go. Its unique approach to transporting and presenting food makes using the Lunch Purse a pleasant dining experience rather than a daily chore.

Sarah Hutley
Industrial Design and Technology

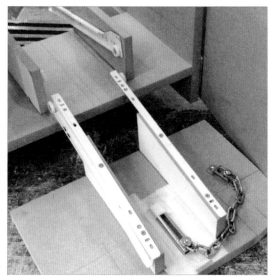
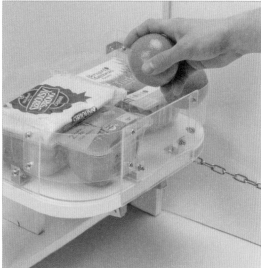
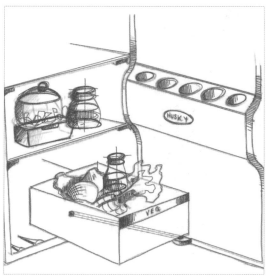

The UK produces approximately 8.3 million tonnes of food waste annually; 5.3 million tonnes of which is considered avoidable waste. The intention of Flow is to help reduce the amount of food wasted each year, by following a user centred approach and focusing on the domestic refrigerator as a key-contributing factor. By introducing Flow's technology into the refrigerator, the lower drawer is extended and raised towards the owner during every interaction. Visual movement highlights the area that stores 'high waste' food items, in turn emphasising the need for their consumption. Additionally, Flow improves not only visibility but also the owner's attitude towards food waste.

Cameron Henderson

Industrial Design and Technology

Lil'Cook

Appliance that encourages children's involvement in food preparation

There is a growing lack of culinary skills and healthy eating awareness among young people in the UK. Research has identified that this is a result of not involving children in the preparation of family meals from an early age, due to safety concerns and lack of time. Lil'Cook is a portable induction cooker designed to address these challenges. The cooker is equipped with a selection of specially designed cooking vessels that allow children as young as six to learn to prepare a variety of nutritious savoury meals (with supervision). The vessels use induction heating technology and insulating silicone rubber to optimise safety and minimise burn hazards, ensuring that the external temperature of the appliance does not exceed 48°C during use.

Adam Wadey

Product Design Engineering

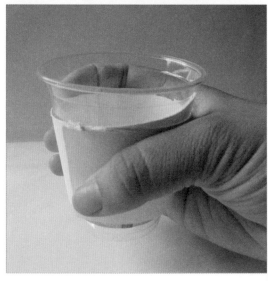

This packaging format project is in collaboration with a reputable, international Fast Moving Consumer Goods (FMCG) company. The consumable packaging format is specially designed for people who regularly snack healthily and want to try something new. Most consumers have proactive lifestyles and often commute using their own vehicles. There are many common accidents from spillages or even burns from their bought snacks or drinks whilst on the move. After many RIG tests and design development these innovative packaging format concepts consist of combinations of thermal insulating material and clever material engineering.

Winnie K.Y. Tang
Product Design

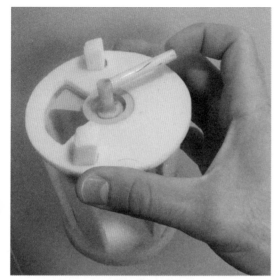

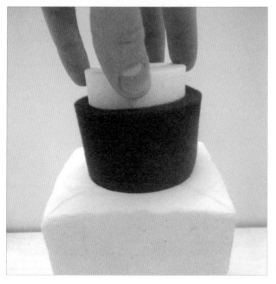

With meal consumption fragmented into more frequent smaller meals and taken in more diverse locations, on-the-go eating is becoming increasingly popular in the UK. In collaboration with Heinz UK, this project aims to develop a disposable drink package for in-car dining. Designed primarily to fit the cup holder of cars, the product contains an attached foam sleeve which provides versatility to fit into a variety of cup holders at the same time offering the user with hand insulation when used with a hot drink. Unlike existing drink packages, the product reduces the chance of spillages and contains a variety of innovative features for optimising usability and user interaction.

Exequiel Di Salvo

Industrial Design and Technology

Hot Pax

A portable food heating device

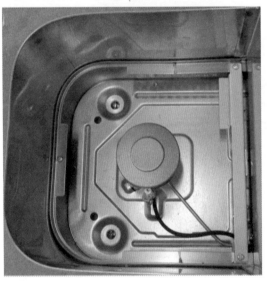

The project is to develop a portable device capable of heating pre-cooked food or snacks from an initial temperature to a predetermined temperature for an extended period of time. The product is designed for people who do not have the time or facilities to heat food or snacks in a microwave oven or access to mains electricity. The portable food-heating device is a perfect accessory to any picnic, beach, dormitory, campsite, or function where you will be on-site for long periods of time.

Yunqian Du

Product Design

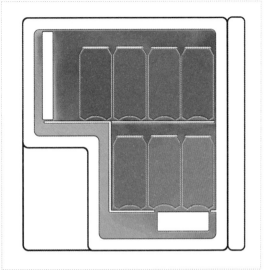

Working in collaboration with Husky Retail Ltd, the market leaders in drinks refrigeration, the design of a counter-top drinks fridge was reviewed with the aim of incorporating new design features. Through primary research a potential need for even cooling and the fast chilling of beverages was highlighted. By following a rigorous design process, technical details were resolved whilst always considering user needs and requirements. Fluid dynamics and computer simulations were employed to prove experiments and validate design modifications. Physical prototyping and human-centred design techniques were then carried out to affirm the key design decisions.

Chris Naylor
Product Design

G&T Jeeves

Automate your merriment!

Gin and Tonic drinkers may be regarded as sophisticated and charismatic, but with every drink the ability to pour the next is compromised. Fortunately, G&T Jeeves is here to help. Through an intuitive interface - click a few buttons to choose your measures - your perfect G&T arrives in your hand. Jeeves is controlled with a custom designed motor driver programmed with C code.

Two high torque DC motors turn him, and a series of sensors ensure he always knows what drink is where, even if you shut him off mid pour. He even precisely times the pouring of your drink to prevent over fizzing and subsequent flattening. Love G&T, Love autonomy, Love G&T Jeeves! (Bow Tie not included)

Cianan O'Dowd

Industrial Design and Technology

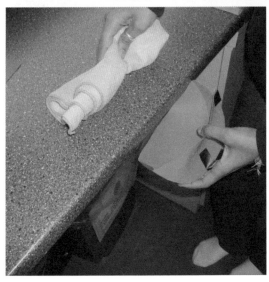

In 2011, almost 84,560 cases of food poisoning were reported in the UK. This project was an investigation towards maintaining cleaner environments in commercial kitchens during service hours. Existing cleaning processes and products were studied, as well as food hygiene perceptions among kitchen staff and a video ethnography was undertaken, all providing insights into possible innovative solutions. On-the-Go Cleaning Unit is an easily-accessible cleaning pack, with appropriate cleaning products to help keep work-top surfaces, a common breeding surface for contaminants, cleaner. The Unit will help to streamline the cleaning process, thus reducing cleaning time, making cleaning compliance easier and improving behaviour.

Peter Cheung
Industrial Design and Technology

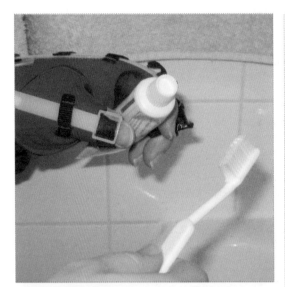

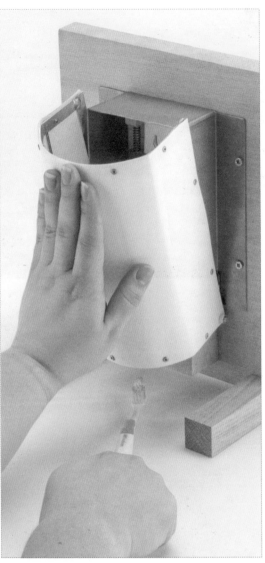

In the UK, Rheumatoid Arthritis (RA) is thought to affect over 690,000 adults. It is a progressive chronic disease causing inflammation of joints which can affect the whole body, particularly the hands and wrists. Sufferers experience difficulties holding and manipulating their toothbrushes while struggling to squeeze the toothpaste out of its tube before cleaning their teeth. This design encapsulates an inclusive solution to the dispensing of toothpaste without the conventional need to squeeze the tube by hand, removing the painful actions familiar to RA sufferers. The toothpaste tube is located and stored securely within the product's shell. By pressing against the ergonomically designed front, the desired amount of toothpaste is dispensed onto a waiting toothbrush.

Deborah Wrighton
Industrial Design and Technology

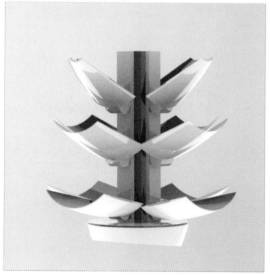

PineCone concept designed for people with osteoarthritis of the hand, a condition that currently affects over 5 million people in the UK. Since childhood, our hands have explored shapes, objects and textures. Unlike a basic heat pack, or a specific orthopaedic hand exercising device, this design concept allows someone to gently discover the textured surfaces on the product. Together with a central, reusable heat pack, they can softly exercise their joints, stimulating the synovial fluid and increasing blood flow. The product allows the user to experience home therapy to alleviate their condition in a relaxed and natural manner.

Matt Baldwin

Industrial Design and Technology

Forget Me Not

Communication models to map emotional changes in older adults

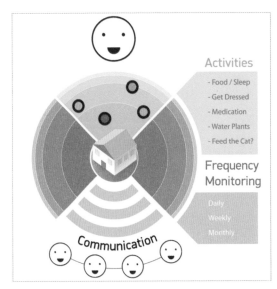

The decline in social relations of older adults has a direct effect on well-being in later stages of life as was highlighted by studies from Age UK and the: "Keeping Connected Challenge" by the Design Council. Often symptoms that signal the start of critical conditions pass unnoticed until they reach a clinical stage. By passively identifying subtle changes in daily life, prevention could occur by way of social intervention. This project involves the creation of a remote sensor network based on interactive household devices that will promote a more gentle and indirect way to communicate emotional changes to a circle of important recipients.

Dimitrios Stamatis

Industrial Design and Technology

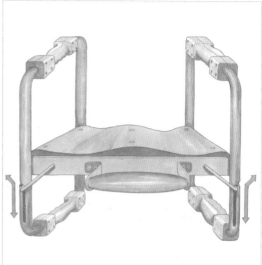

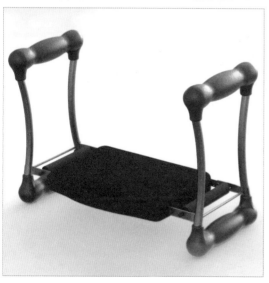

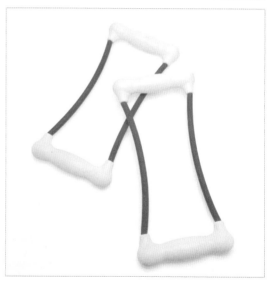

Recent figures state that in the UK alone, over two billion pounds are spent in garden centres every year. The current basic garden kneeler has been a million pound selling product for PBFA and Marks & Spencer in previous years, however its design has since become quite dated. The garden equipment sector that the kneeler resides in has great potential in relation to changing demographics, their interest in gardening and the profile of the M&S customer. This is the basis for the study. The product has been taken back to its roots and has undergone a fresh and dynamic re-design, interweaving improved aesthetic values with crucial inclusive and semantic design aspects that have previously been overlooked.

Phil Verheul

Industrial Design and Technology

Core
Supportive apple picking

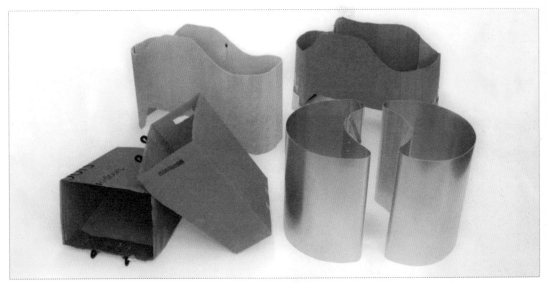

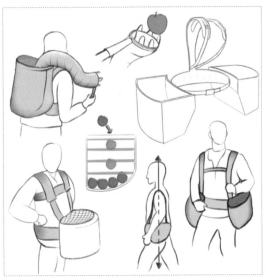

Through an investigation of current apple harvesting practices, it is clear that existing equipment is largely inefficient, leading to strain for pickers, as well as damage to the fruit. Existing solutions provide little support to counteract these problems, resulting in a high probability of injury and discomfort. Supportive apple picking is a project that re-evaluates the apple picking process, with a strong focus on usability.

Core is a twin-bucket system that applies a balanced load onto the vertical spinal axis, greatly improving posture and comfort. This new arrangement also improves the carrying capacity and reduces the risk of apple bruising. Core has been proven to provide an improvement in the overall efficiency of the picking process, as confirmed through user testing.

Dan Cherry

Product Design

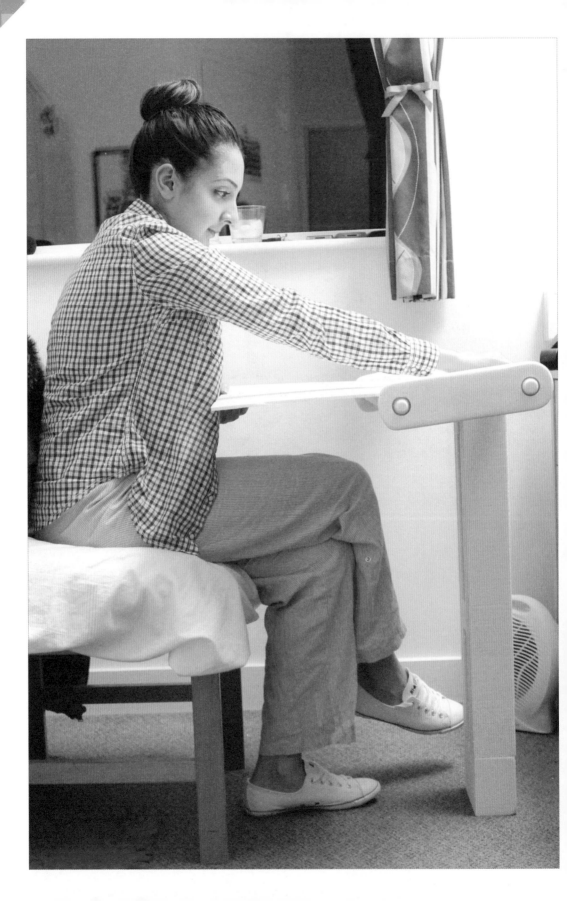

The Vertebrae Desk

An adjustable desk to help improve posture

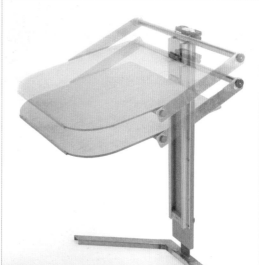

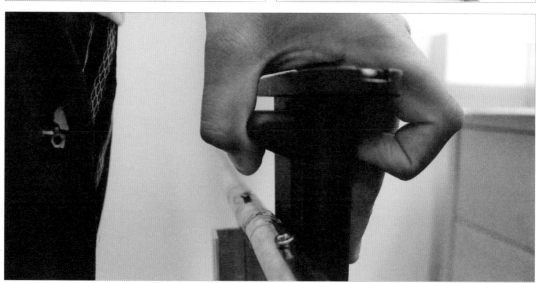

A desk is a vital piece of furniture that aids learning in all school years. To control lower back pain, good posture should be established in childhood. In collaboration with FIRA, the project intended to provide a better solution to existing static desks, which currently influence uncomfortable postures with negative health implications.

The Vertebrae Desk uses an intuitive and fast mechanism, enabling the user to adjust the desk to a suitable height. The stackable structure enables accessibility and space. A 15° desk slope provides the correct reading and writing angle to attain an upright posture, encouraging an optimistic learning environment.

Sukhi Assee

Industrial Design and Technology

An Intuitive Task Chair

An understandable method of adjustment

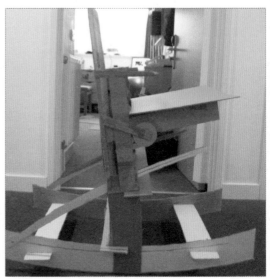

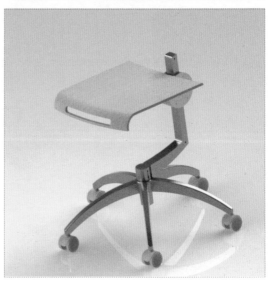

The key purpose of a task chair is to provide a healthy posture and maintain consistency in everyday seating needs in the work environment. People working in flexible regimes require a chair for irregular periods of time and often share unconventional workspaces for tasks, leading to frequent, unhealthy seat height adjustment. This project takes a holistic approach to adjustment, starting with the most important and basic adjustment feature. Re-thinking the current gas-lift system produced an understandable, sustainable and viable solution. The proposal seeks to challenge existing conventions by focusing on the context of flexible working.

Luke Bacon

Product Design

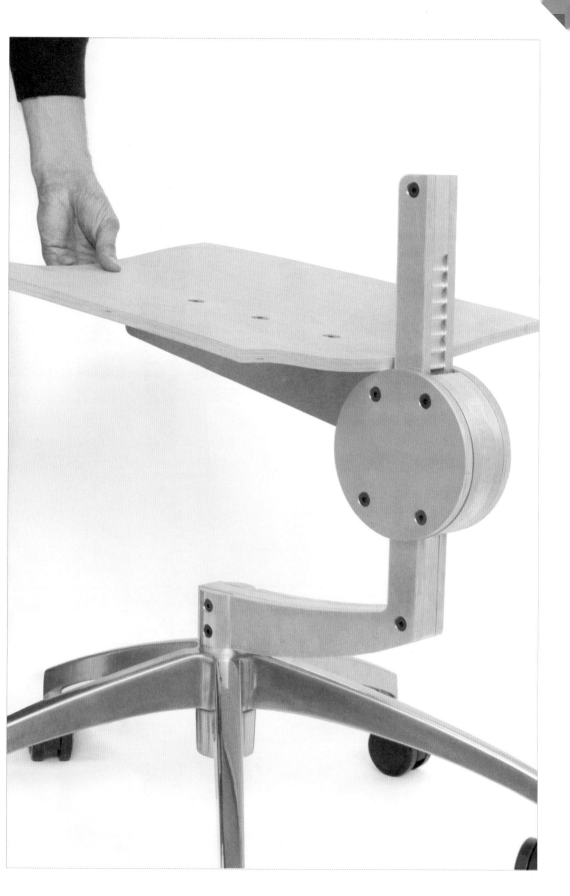

When people develop hand disabilities, they often need to buy completely new items or use specialised products. This project focuses on creating eating and drinking products for people with reduced hand mobility and strength. The CorkAid cutlery grip adapts to anyone's cutlery and a cup sleeve that fits around regular coffee cups. A cork material was specifically developed to help in gripping while taking advantage of cork's natural qualities. CorkAids allow people with less hand dexterity to carry on using the same items as others. The project realisation was made possible by taking influences from the personal experiences of people with reduced hand mobility, and by researching, developing and optimising the natural qualities of cork.

Alec James
Product Design

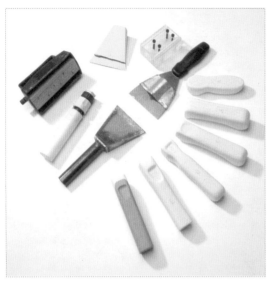

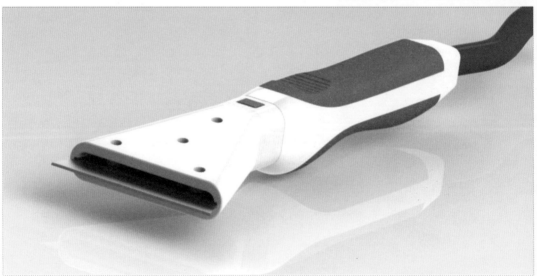

Removing wallpaper is a long and daunting task that both professionals and non-professionals would rather avoid. There are a variety of products and processes available for consumers to choose from, but the current solutions are time consuming, awkward and require too many stages to do the job correctly. With the DIY market expected to be worth a staggering £11.4 billion by 2015, and considering that wallpaper strippers have been proven to be a consistently high seller for many years, the lack of innovation within this market sector is surprising. Stream is a multifunctional tool that is user-focused, combines different stages of the steam removal process and makes the daunting task of wallpaper removal more efficient and approachable.

David Crittenden
Industrial Design and Technology

Hearing Protection and Communications for Soldiers

Ergonomic redesign of hearing protection and communication system

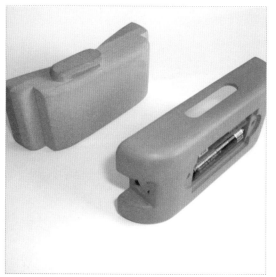

This product was designed in collaboration with Racal Acoustics, and aims to help protect frontline soldiers on operations helping to reduce the number of casualties with extreme hearing damage. By improving both ergonomics and comfort it will encourage more soldiers to wear the equipment properly. By improving the interaction with the device the hope is to also reduce the cognitive load and help them to concentrate on their mission rather than the equipment.

Jonathan Hyslop

Product Design Engineering

Work Space

The design of a desk for the limited space within homes

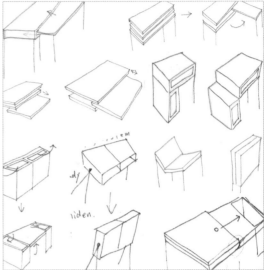

1.3 million people work from home and there are 3.7 million employees who sometimes work from home or use home as a base in the UK. However, home space is getting smaller especially in urban areas and most homes do not have a spare room for an office. The product is a desk which is designed for the limited space. The desk is designed based on analysing the space at home, people's working habits and space management of the desk. Since space is limited, it has to be as small as possible. Having the slider desk allows extra space and more efficient use of desk by moving a laptop away from you and towards you.

Moyo Ralleigh

Product Design

The main aim of the creation process during this project to design a product which will optimise the experience in retail environments pregnant women. With the use of technology and new materials, this product will be inclusive and innovative. Most importantly, it will help to achieve better communication both with retailers and other customers, adding even more value. The focus point of the project involves the post shopping process and how the purchased items are being transported from the retail environment to the desired destination. The final output of the design process will greatly decrease the strain of carrying shopping for pregnant women.

Polina Liarostathi
Integrated Product Design

Collaborations

Designplus facilitates collaborative Major Projects with a wide variety of partners from entrepreneurial startups, charities, SMEs to large multi-national companies. We apply our knowledge and skills to professional project briefs, receiving expert advice and support throughout the project from our clients.

Sony Europe

Sony VIBE

Guy's and St Thomas' Hospital

Patient Information Panel

Oral Development Pack

Infant Oral Motor Development Toy

OCT Unit Workstation

Nasogastric Tube Fitting

Starling Lunch Carrier and Accessories

Gray Starling Ltd

Anti-dive Suspension

USE

Seat Post

Professor Alison McConnell

Rehabilitation Walking Aid

Hot Pax

Seafield Pedigrees Ltd

Blackwood Foundation

Forget Me Not

STAT - Tracking

u-blox

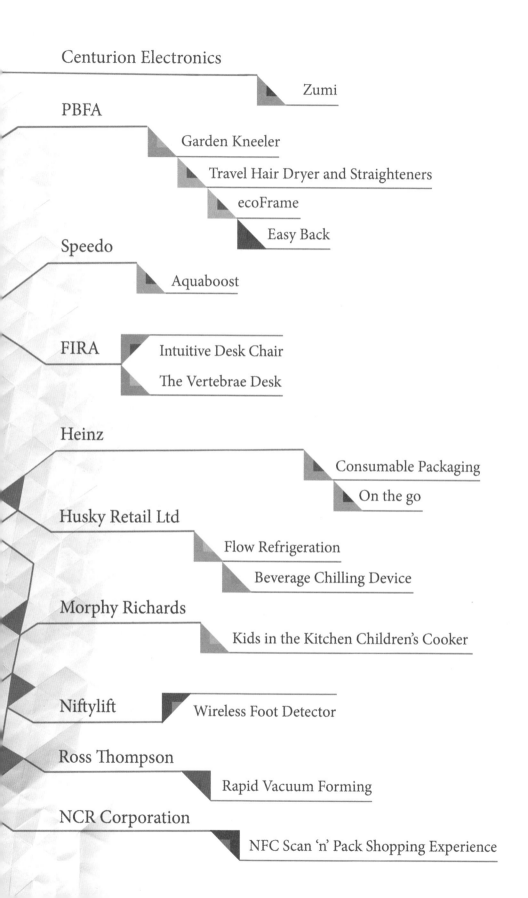

Centurion Electronics

Zumi

PBFA

Garden Kneeler

Travel Hair Dryer and Straighteners

ecoFrame

Easy Back

Speedo

Aquaboost

FIRA

Intuitive Desk Chair

The Vertebrae Desk

Heinz

Consumable Packaging

On the go

Husky Retail Ltd

Flow Refrigeration

Beverage Chilling Device

Morphy Richards

Kids in the Kitchen Children's Cooker

Niftylift

Wireless Foot Detector

Ross Thompson

Rapid Vacuum Forming

NCR Corporation

NFC Scan 'n' Pack Shopping Experience

Buddy

Administering the antidote to an opiate overdose

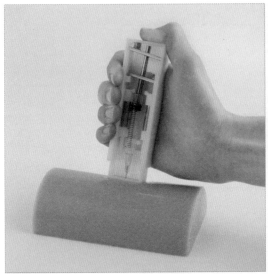

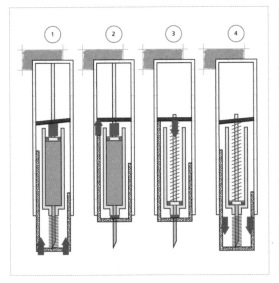

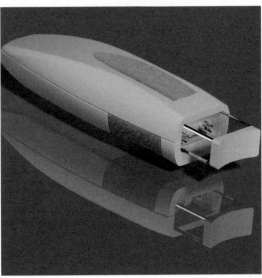

Overdose is now the largest cause of death amongst injecting heroin users. The NHS is currently running drugs trials, allowing carers of heroin addicts to carry the antidote to a heroin overdose and training them to safely inject the drug using a needle and syringe for use in an emergency situation. This product has been designed and developed to inject the drug without the need for medical training. If the trials are successful, the NHS can prescribe a usable and intuitive device to carers that will save lives without having the need to individually train each carrier how to use the device.

Sophie Randles

Product Design

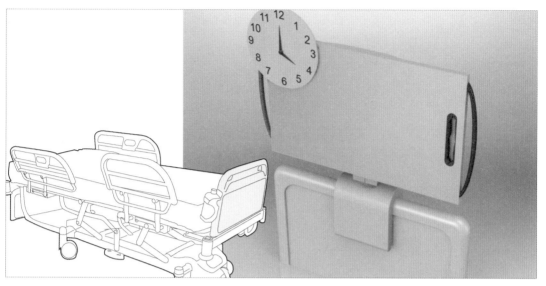

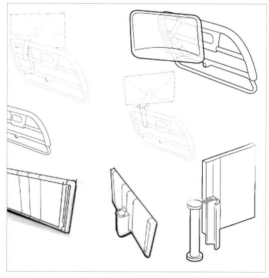

In collaboration with Guy's and St Thomas' Hospital, this information panel is a development of a board being trialled at St Thomas' Hospital. It is used to display pictures, messages, and other useful information for the patient and staff in intensive care. The aim of the board is to improve patient communication and reduce patient stress. My Board is a development of the prototype panel used by St Thomas' Hospital. With its book like features, the panel can be easily fastened to the bed, but just as easily removed when the entire bed space is needed. Maintaining the same function as the prototype, the new concept, is thinner, lighter, fits subtly into the environment, and is more intuitive to everyone who come into contact with it.

Sunil Patel

Industrial Design and Technology

Kulinda

Protecting against HIV transmission in breastfeeding

22.5 million people are living with HIV/AIDS in sub-Saharan Africa and 90% of the 370,000 annual infections in children are avoidable, caused by mother-to child transmission with 40% of these being through breastfeeding. Flash Heating is a process that enables HIV positive mothers with limited resources in sub-Saharan Africa to treat their breast milk before feeding to their babies, providing them with food and nutrients in a safe, economical way. Each Kulinda costs 20p and can also be provided with a pictorial instruction booklet, bag and container. Kulinda improves the reliability of the Flash Heating process with a simple, clear indicator to show when the milk is safe, reassuringly preventing mother-to-child transmission of HIV.

Emily Riggs

Product Design

SNUG

A securing device for NG tubes

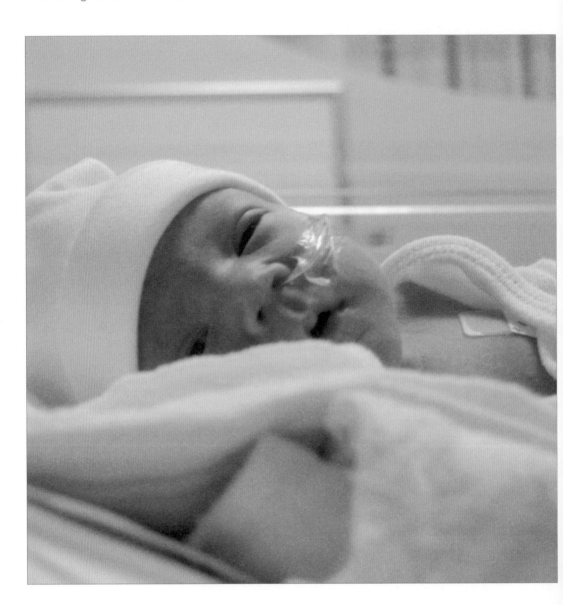

In the past 40 years nasogastric (NG) tubes have been used to provide total or partial nutrition for many infants, with the NHS purchasing approximately 1,000,000 NG tubes annually. The prominent problem with NG tubes in infants is their unprescribed removal, often occurring several times a day. A result of this is additional discomfort for the patient and increased costs for the NHS. In collaboration with Guy's and St Thomas' Hospital, SNUG was developed to help both specialists and parents secure NG tubes, using an easy two-step method. SNUG uses new fixing techniques and a newly designed shape to significantly decrease the number of untimely removals. This is combined with a specific focus on ease of use and aesthetic qualities that minimise the appearance of the tube.

Laura Hodges

Product Design

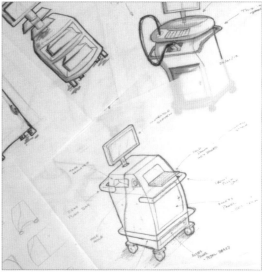

This project was managed in collaboration with Guy's and St Thomas' Hospital to design an Optical Coherence Tomography (OCT) Unit Workstation Trolley. The trolley will house OCT equipment, used in a pioneering technique to help recognise and diagnose Melanoma Skin Cancer. This equipment provides an image of the scanned area immediately, a massive improvement on the current diagnosis time of three months. The trolley design proposal will help to bring this new technology into a wider market and speed up the long, complicated process of Melanoma diagnosis and treatment.

Tim Logg
Industrial Design and Technology

In-Vivo Piezoelectric Energy Harvesting Device

Harvesting energy from the motion of a human walking

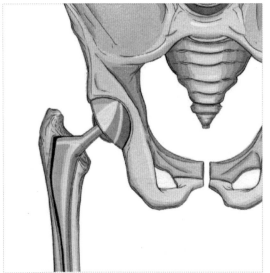
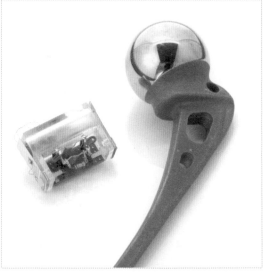

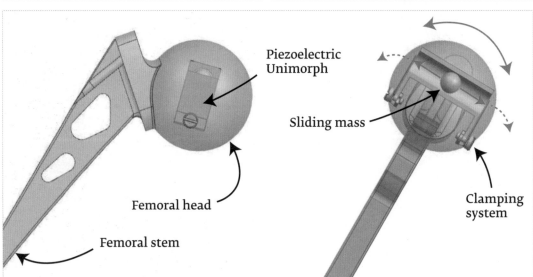

Piezoelectric Unimorph

Sliding mass

Femoral head

Femoral stem

Clamping system

A device was designed and built that could successfully harvest energy from the rotary motion experienced by the femoral head of an implanted hip prosthesis. Spherical missiles roll along the shaft of the device as the femur swings about the hip striking piezoelectric cantilever plates at either end. Stress induced by these impacts creates an electrical charge that is stored and can be used to power a strain gauge and transmitter allowing transmission of medical information from the prosthesis. This information can help detect the onset of aseptic loosening in the prosthesis allowing measures to be taken to prevent Osteolysis.

Luke Kavanagh

Mechanical Engineering

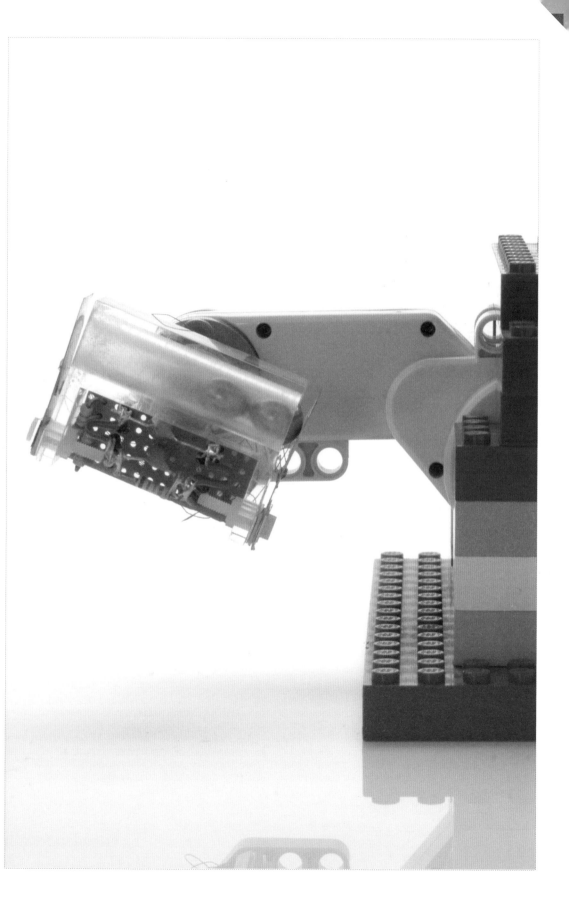

CFD Study of Flow in a Realistic Aneurysm Model

Analysis of cerebral aneurysm blood flow RYSM model

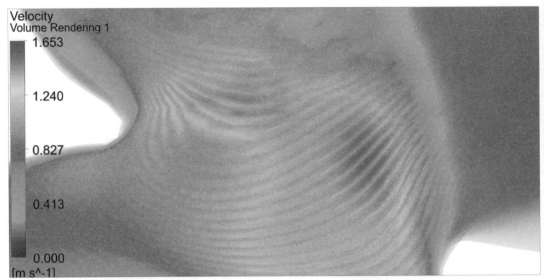

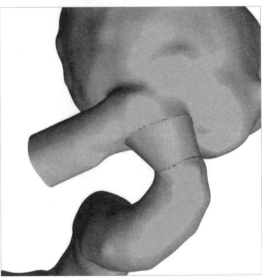

As the leading cause of subarachnoid haemorrhage, it is clear that the study of intracranial aneurysm behaviour is of paramount importance in efforts to improve its management. This study looked at simulations to predict flow patterns in a giant internal carotid artery aneurysm which had a constriction (stenosis) just proximal to the aneurysm ostium. Analysis showed that the concentrated flow jet created by stenosis constrictions drastically increases sub aneurysmal haemodynamic attributes, in some cases nearly doubling these parameters as compared to flows without stenosis. The resulting flow jets serve as a likely candidate for aneurysm growth, and may even be a key element in aneurysm formation.

Hardeep S Kalsi

Aerospace Engineering

Elevator Pro
Root and erupted teeth extraction device

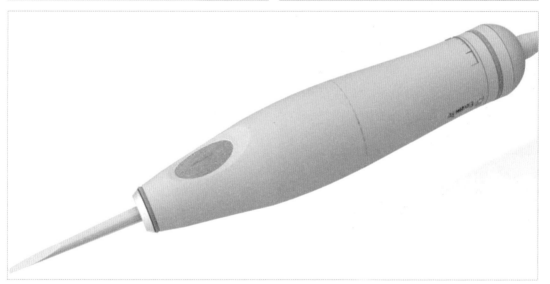

The removal of roots and wisdom teeth are extremely traumatic for patients. Dentists do not feel confident and relaxed during the extraction process. Existing removal tools injure the jaw bone. They also carry high risk of injuries to other parts of the mouth. These issues are caused by applying high unknown forces towards the tiny area of the socket during the process of losing the root. The Elevator Pro was developed in collaboration with an experienced professional dentist. The design solution enables dentists to adjust the right amount of force for different bone densities. The movement of Elevator Pro's blade is automated and is initiated by an electrical switch, instead of manual force by the dentist.

Shima Roozbahani
Product Design Engineering

Pace

Effectively reducing rehabilitation time post-knee surgery

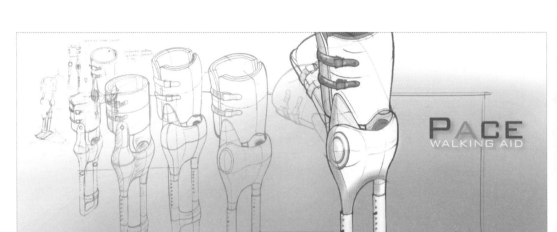

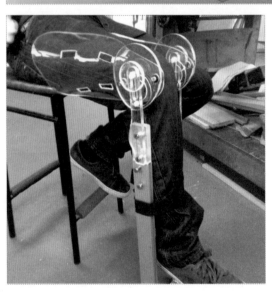

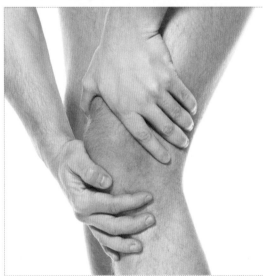

More than 1.7 million people suffer from meniscal tears each year. Meniscus tear is a common sports injury and is especially prevalent among competitive athletes in football, rugby and basketball. Patients who have partial or total meniscectomy have an increased risk of developing osteoarthritis over the following four to eight years. This project focused on the design and development of a medical aid for post-knee surgery patients suffering from meniscal tears, which takes the weight off of the knee and allows the patient to walk 'normally', while removing any weight/compression from being transferred through the knee joint.

David Johnston

Industrial Design and Technology

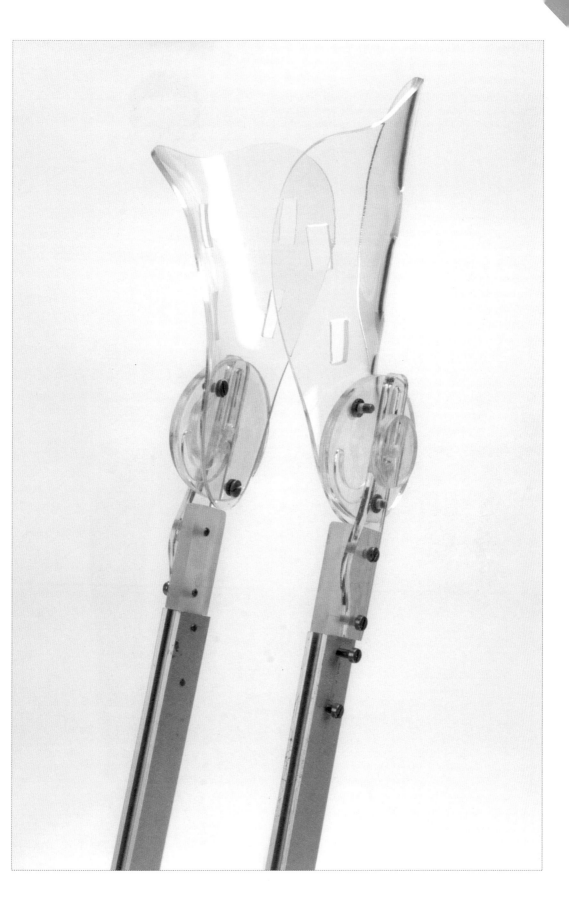

Creative Itinerary

We have been in Brunel for a long time, and it has been a long journey. Some things we learnt were taught, some we learnt through doing, and some through making mistakes. This avatar symbolises us as a student body and represents our thinking, our process and our journeys.

Mouth

Communicate, collaborate.

Watch

Never miss a deadline.

MADE IN BRUNEL

Mind

Question everything, discover problems.

Eyes

Look for inspiration, fuel your creativity.

Heart

Do what you love, love what you do.

Gut

Go with your instinct.

Hands

Make models, make coffee, make stuff happen.

Feet

The long journey starts with the first step.

Aquastop
Eliminating water ingestion during swimming

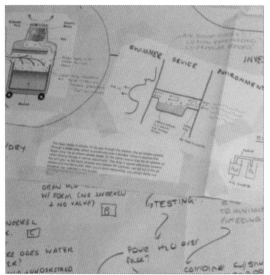
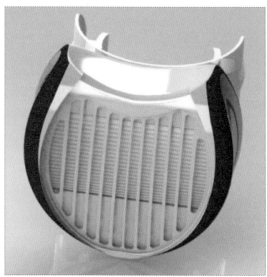

Open water swimmers train in rough water environments and therefore swallow a lot of water when breathing. This high level of water ingestion means that they are at high risk of contracting waterborne gastroenteritis. Although snorkels can prevent the ingestion of water when swimming in open water, they prevent the swimmer from developing a natural breathing technique, as they no longer have to lift their head out of the water to breathe. Aquastop prevents the ingestion of water but allows the swimmer to develop a natural breathing technique during training. The hydrophobic filter repels water, but allows air to flow freely through, therefore preventing the ingestion of water while allowing the swimmer to breathe comfortably.

Victoria Gibson-Robinson
Product Design

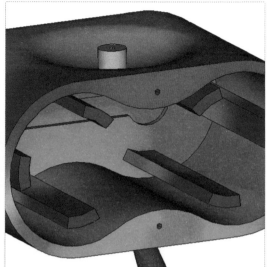

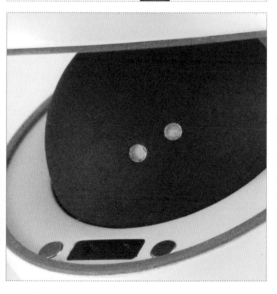

On average, two to five minutes are spent warming up a squash ball by hitting it hard against the wall 50 to 100 times before playing a game. Not only does this waste court time, but it can also increase risk of muscle strain while the body is not fully warmed up. To many players this is a very tedious activity that has to be undertaken before playing.

The Squash Ball Pre-warmer focuses on effectively heating squash balls before a game so that they are ready to be used as soon as the court is entered. By developing specific heating elements, the product will heat a ball to 45°C within three minutes while also providing heat to a secondary ball in case the first one bursts.

Louis Garner
Product Design Engineering

Revolve Bicycle Seat Post

In collaboration with Ultimate Sports Engineering

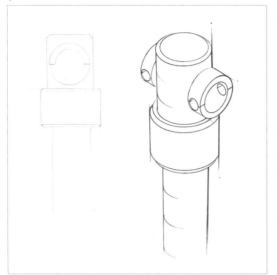

One of the few remaining industries experiencing growth, the cycle industry is booming. A new breed of rider appreciates performance whilst promoting smooth ride-quality and simplicity of use. In this performance environment, bicycle jewellery that brings functionality beyond that of existing products wins the heart of cyclists and helps them towards being competitive. The Revolve concept turns the current application of small, fiddly bolts on its head, bringing increased performance and usability by combining the functions of several components into one simple action. Using a large diameter collar, force is spread across a large area, aligning with the properties of the material to create a product of the lowest weight, hugely important in a sport driven by performance.

Samuel Edwards

Product Design

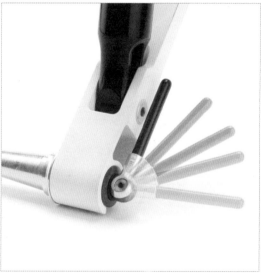

The final product, S.U.B., stands for Stability Under Braking. With its innovative, unique, and patented anti-dive technology it aims to make the rider more confident and comfortable when riding off-road. Using the braking force itself, through levers and pivots, it is able to stop dive and improve the ride. The single-sided design makes it easier to change a tyre or inner tube whilst also being stiffer and lighter than conventional suspension units. The ball-bearing pivot joints and quick release mechanism ensure the anti-dive front suspension unit is on current cycling trends. Finite element analysis design tools and the use of composite materials ensure it has the highest strength to weight ratio.

Roland Skinner
Industrial Design and Technology

Ski Boot Walking Aids

Attachable pads to improve the comfort of walking in ski boots

In a ski resort a skier will have to walk long distances from their hotel to the lift in the morning and back in the afternoon, as well as around restaurants and the resort, all while wearing their ski boots. Ski boots are constructed of a rigid polyether shell to maximise performance for skiing, however this makes them difficult to put on and take off, and also offers no flexibility in the sole. This product can be quickly fixed to the soles of ski boots to improve comfort and walking posture, and then quickly removed and stored at the back of the boot while skiing.

Rob Dowling

Industrial Design and Technology

SecureSki - Portable Ski Security

Compact security solution allowing skiers to easily secure their skis

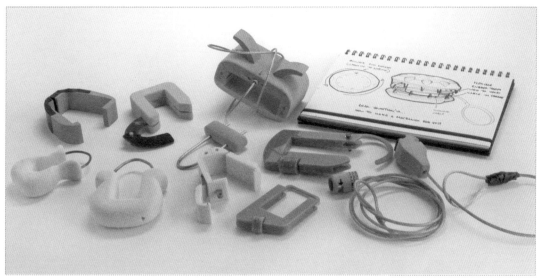

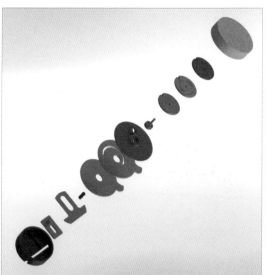

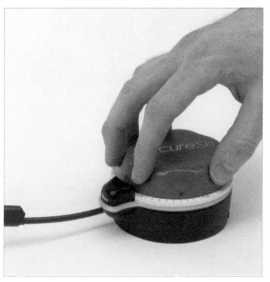

It is estimated that 80 million skiers visit one of the world's major ski resorts each year to take part in the sport they love. Every season over 30,000 skis are reported stolen or missing. With ski hardware costing in excess of £3000, having one's skis stolen can be a very stressful and costly experience, particuarly with insurance companies often making it extremely difficult to claim for their theft. A simple solution is to secure your skis whenever they are left unattended. However, current ski locking solutions are not adequate in all areas. They are rarely intuitive or pleasant to use and often sacrifice security for portability. The SecureSki sets out to address the key factors with equal weighting to create a secure, highly portable and enjoyable product to use.

Salvatore Bella

Industrial Design and Technology

Climbing is a rapidly growing sport worldwide; attracting people with a zest for life and a wish to face new challenges. The most popular form of the sport is 'traditional' climbing, where climbers place their own protective devices into the rock to prevent a dangerous or potentially lethal fall. A common characteristic of the terrain is that rock cracks flare outwards and become more open when many of the current climbing protection devices are placed in them. This creates an increased problem that raises the chances of sustaining serious injuries or even death from a fall. The objective is to provide a means of protecting these rock features, together with the capabilities of existing climbing protection devices.

Ben Crichton
Industrial Design and Technology

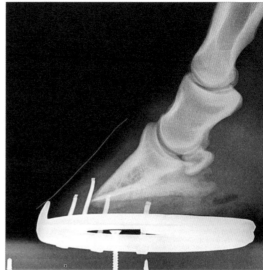

Horse racing has been a popular sport for centuries; it has been driven by speed and power which have led to breeders and trainers aspiring to produce the ultimate racehorse. However, with one in ten thoroughbred horses suffering from an orthopedic problem, this project aims to re-evaluate the way in which horseshoes are used in the racing industry. The new concept is designed to imitate the natural expansion of the hoof while retaining the benefits of grip and protection. By accommodating natural shock absorption, the stresses and strains that a horse may endure can be transferred evenly through the body and reduce the risk of injury.

Barnaby Hunter
Product Design Engineering

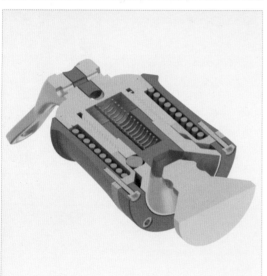

Trapezing is an exhilarating technique for balancing high performance sailing dinghies against the force of the wind. Current harnesses use a crude hook and ring system, which can snare on ropes or boat fittings during a capsize. This causes 30% of capsize entrapments, and has tragically led to seven fatalities since 2002. Experienced sailors have been consulted during the development process to identify flaws in existing quick release systems and ensure the new device meets the needs of the most demanding users. Off the Hook is an innovative new approach to the problem. The patent pending mechanism and harness attachment make trapeze sailing faster, safer and more intuitive. Designed and engineered to be dependable in the most extreme marine conditions.

Simon McNamee

Product Design Engineering

Structural Analysis of a Badminton Racket

Using Finite Element Analysis in racket development

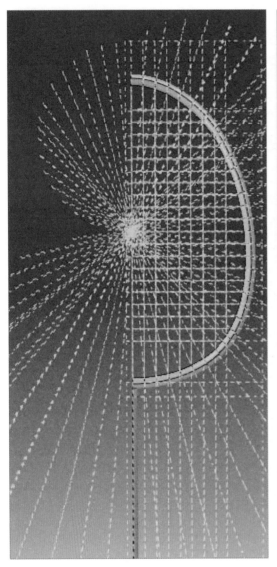

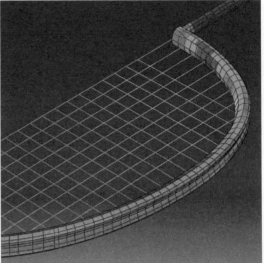

Badminton is a popular sport worldwide, with 173 members being represented in the sport's governing body, the Badminton World Federation. From its origins, badminton has evolved into one of the fastest, most physically demanding, racket sports in the world. Manufacturers are constantly looking to improve rackets in order to provide players with an edge on the court but little previous research has looked into using Finite Element Analysis (FEA) as a method of racket development. This project creates a basic FEA model of a racket which, when compared to experimental data, accurately represents the forces at work in a racket under the load of a shuttlecock. This can be used for further development of racket technologies.

Christopher Luscombe

Mechanical Engineering

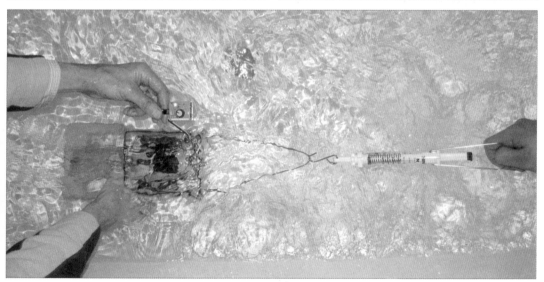

In England there are 5.6 million swimmers. Competitive swimmers are constantly looking for ways to improve speed and technique through the use of cutting-edge technologies. Aquaboost is a Personal Water Propulsion Device making assisted and resisted training styles easy to implement. Studies have found that resistive-sprint training is effective at building muscle strength whilst assistive training leads to an increased stroke rate and helps to improve hydrodynamic position. However, the range of products available to provide this sort of training is limited and most have significant drawbacks. Worn around the waist, Aquaboost creates additional thrust helping familiarise swimmers bodies with 'Race Pace' and increasing strength.

Emily Menzies
Product Design

Smart Wetsuit
Bio sign monitor for surfers

The search for new waves never stops. This is drawing surfers to some of the remotest locations around the world, including some of the coldest. The smart wetsuit device can measure your core body temperature, and then warn you when it starts to drop, helping to prevent hypothermia. The device also contains a GPS module, that records the surfer's telemetry to be shared with an online community.

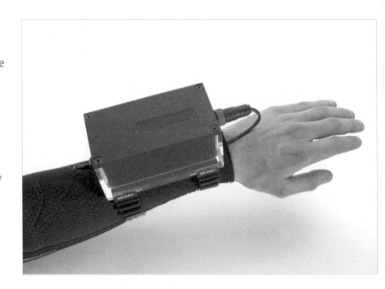

Mark Taylor
Product Design Engineering

Easy Adjust Dive Mask
Maximising the comfort and reducing the strain felt by divers

With over 30 different chains of Diving Qualification companies operating globally the, once relatively specialized, sport is on the increase, with equipment also improving dramatically over the last decade. However one part that has stayed relatively untouched is the diving mask itself. The objective of this project is to create a one-piece attachable strap that will be compatible with most existing masks allowing for precise and incremental changes in the levels of tightness to be made in an ergonomic and comfortable way, reducing the stress and maximizing the overall comfort of the user.

Clayton Jack Ellacott
Industrial Design and Technology

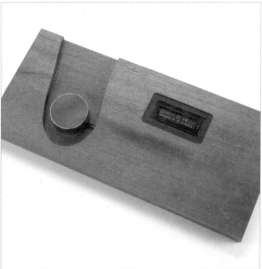

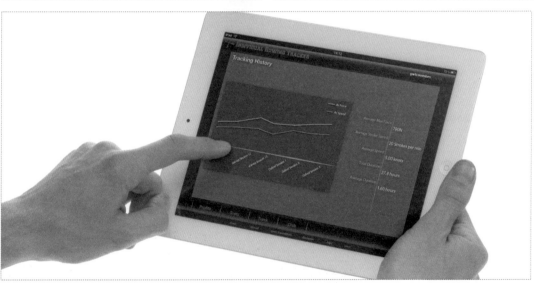

Many athletes are using advanced technology to aid their development. So far in rowing, technology has only been easily available for the world class and Olympic athletes; smaller clubs and individuals can very rarely use such devices to analyse their performance. The individual rower tracking device is an affordable way to aid all rowers, using technology to track and give feedback on their strength and speed. The data collected by the device can then be uploaded and analysed using a computer or smart phone.

George Coombes
Product Design Engineering

PIC Darts

Automatic scoring system for darts

PIC Darts uses two layers of conductive foam on the board to form a bridge, the dart acting as a connection between the two layers to register the score. By separating each segment of the board into unique bridge circuits the scoring system is able to add up each player's score depending where they hit and keep a record of their total.

Louis Garner

Product Design Engineering

Snooker Game

Snooker training aid

A training aid positioned for snooker angles and a game. This product will be on snooker and pool tables, for players to practice manoeuvring the cue ball and colours through tight gaps and different angles utilising the cushions. Points are awarded for potting cleanly through each tunnel to allow the player to test themselves against the clock or an opponent.

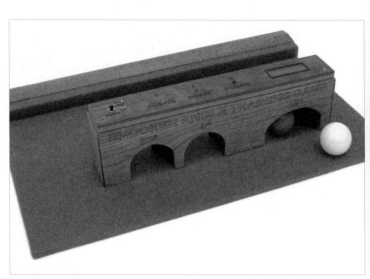

Jamie Phillips

Product Design Engineering

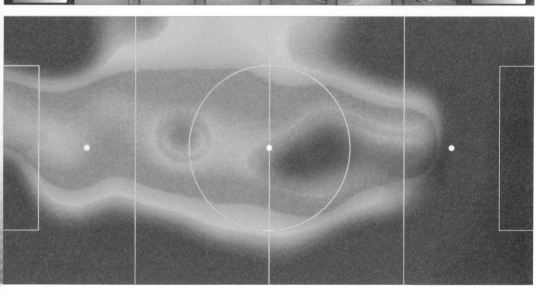

Sports Team Analysis and Tracking application (STAT) is an online platform, which enables match data and highlights of any outdoor football games to be collected and replayed. The data used can be collected from wearable GPS wearable tracking devices making it suitable for hobbyists and amateurs sportsmen. It provides an opportunity for players across all levels to revisit their previous matches, share these with friends, improve teamwork, compare themselves with professional players, follow their improvements and encourages them to take greater part in team sports.

Archit Rakyan

Product Design Engineering

Work-Play Balance

Following copious amounts of work, we quickly realised the importance of relaxation. Finding time to unwind was difficult, but the spare minutes we managed to enjoy were filled with the things we like the most. This infographic shows how we balance our lives.

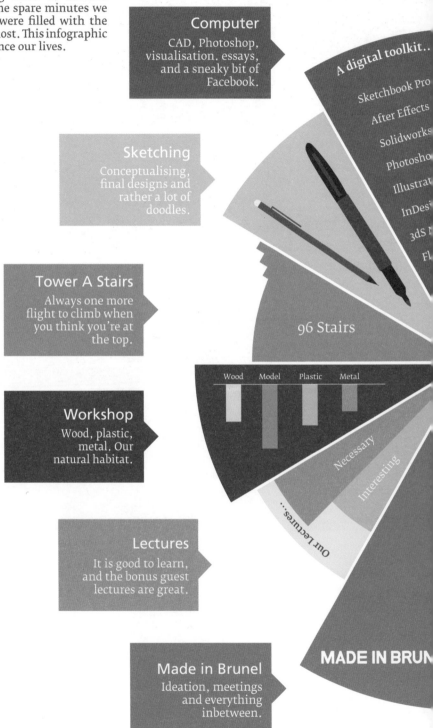

Computer
CAD, Photoshop, visualisation. essays, and a sneaky bit of Facebook.

A digital toolkit..

Sketchbook Pro
After Effects
Solidworks
Photosho
Illustrat
InDes
3ds
Fl

Sketching
Conceptualising, final designs and rather a lot of doodles.

Tower A Stairs
Always one more flight to climb when you think you're at the top.

96 Stairs

Wood   Model   Plastic   Metal

Workshop
Wood, plastic, metal. Our natural habitat.

Necessary
Interesting
Our Lectures...

Lectures
It is good to learn, and the bonus guest lectures are great.

Made in Brunel
Ideation, meetings and everything inbetween.

MADE IN BRUN

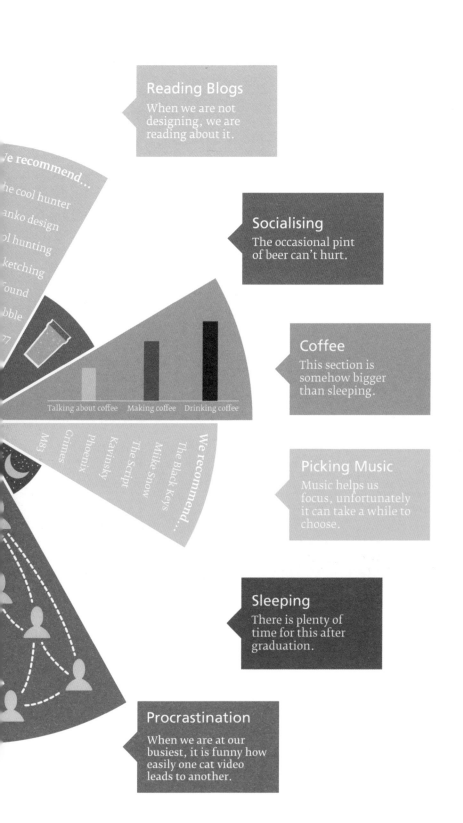

Reading Blogs
When we are not designing, we are reading about it.

We recommend...
The cool hunter
Swissmiss
Booooooom
Design inspiration
FormFiftyFive
Grafik
Creative review

Socialising
The occasional pint of beer can't hurt.

Coffee
This section is somehow bigger than sleeping.

Talking about coffee   Making coffee   Drinking coffee

We recommend...
The Black Keys
Miike Snow
The Script
Kavinsky
Phoenix
Grimes
M83

Picking Music
Music helps us focus, unfortunately it can take a while to choose.

Sleeping
There is plenty of time for this after graduation.

Procrastination
When we are at our busiest, it is funny how easily one cat video leads to another.

John and Rex

Raising awareness of the impact of technology on the elderly

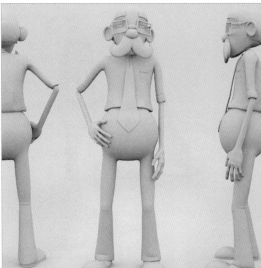

'John and Rex' is a 3D animated short, created to raise awareness of the impact of technology on the elderly. Although the story is some years ahead of the present time it is very likely that in due course the future will see experimental technologies like Rex, a futuristic hover dog, widely adopted. The original idea for John and Rex came from personal experiences of trying to teach my grandparents how to use a computer. It was their struggles with current technology that fuelled the creative process for this project.

Russell Hinton
Multimedia Technology and Design

Tagit

Electronic tag game

In Tagit, the aim of the game is to get the lowest score possible, with each player trying to 'tag' targets on their opponent's backs as many times as possible within a given time limit. Through embedded RFID technology, each tag accumulates a score and records who tagged the user with the information being displayed at the end of the game.

Luke Gray

Industrial Design and Technology

DorDor

Electronic musical chair game

DorDor is inspired by the traditional musical chair game, but using boards instead of chairs. When the start button is pressed, the music will play and two colours of red and green LEDs on each board start flashing. It carries on until the music stops, when each player needs to find a board which has a green light to stand on. Anyone who does not find a board or stands on one with the red colour light should leave the game. The game continues in rounds until only one board goes to a green light and the person who stands on that board first will be the winner of the game.

Shima Roozbahani

Product Design Engineering

K-Pull

Audio cable strain relief

Audio and musical instrument cables are subjected to rough treatment and suffer accidental damage primarily due to force being exerted on the cable. This causes damage to the plug and can damage the connected equipment. Faulty cables provide poor connections and loss of tonal quality. The K-Pull concept attaches to all common audio cable types and provides strain relief in order to reduce damage. The concept also provides an accessible grip to aid in disconnection, discouraging the user from removing the plug by pulling the cable. The K-Pull also provides a method of securing the cable when not in use. The K-Pull protects cables against the common causes of damage in both use and in storage, thus prolonging cable lifespan.

James Hellard

Industrial Design and Technology

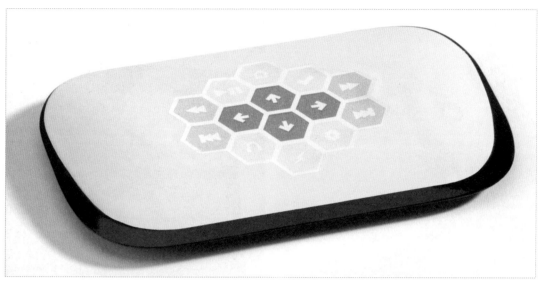

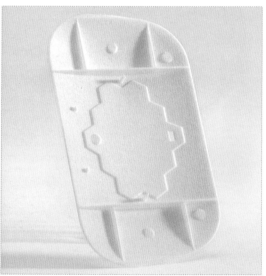

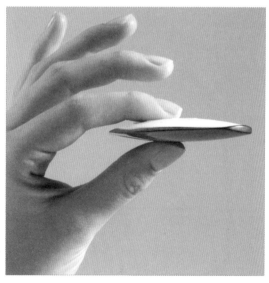

To capture the attention of a child, the remote for an in-car infotainment system has to look and feel visually exciting. By researching new technologies, exploring cues from high end design and by working with tangible 3D form in the model-making workshops an evolutionary journey of design was created from which the product was born.

Care was also spent designing the product for manufacture and assembly; choosing suitable materials that would work with a monocoque chassis and by working through each manufacturing stage, so that no design compromise would have to be made during the downstream design process.

George Williamson

Industrial Design and Technology

133

Vox Box

The portable speaker for outdoor environments

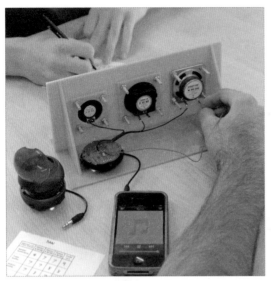

As devices become more portable the environments they encounter demand more from them. Many portable speakers currently on the market are too fragile and delicate for use in outdoor environments. A drop or contact with water will turn the speaker into a useless pile of electronics. VoxBox is the answer to these problems. With four speakers this small Bluetooth speaker is able to direct sound in all directions; great for listening to music with a group of friends. It is also protected against damage from drops and water contact through the use of shock absorbing materials and new water-repellant nano-coating technology. These properties make VoxBox the perfect portable speaker to survive the harsh environments of a festival, beach or skate park.

Andrew Davies

Product Design

Mud Hog

Personal baggage transportation solution for festival-goers

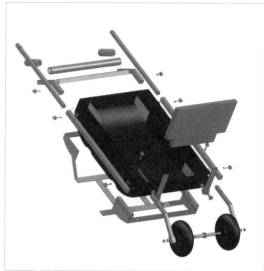

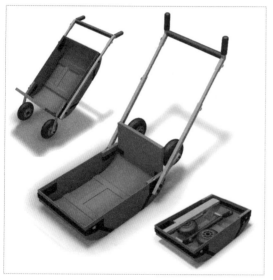

Over 840,000 people camp at festivals each year in the UK alone and festival sites can quickly become very muddy. The terrain conditions of festival sites pose a big problem for festival goers each year, especially when trying to transport baggage across the site. Featuring two different modes, the Mud Hog is a transformable baggage trolley designed with the festival goer in mind. A wheeled mode enables the push/pull of baggage over harder ground and the Sled mode enables easy slide over the softer/slippery ground. The Mud Hog is transformed easily between the two modes, allowing for an easier, stress-free scramble from car park to camp site.

Jon Taylor

Industrial Design and Technology

Drive!

Accelerometer remote control car

Inspired by the tactility of the Wii remote, Drive! is an accelerometer based remote control car that allows the user to direct the car's steering by turning the physical controller. The car itself employs a servo that rotates in proportion to the angle and speed of the accelerometer, allowing the driver to accurately control the car as if they were in the driving seat.

Angela Luk
Product Design Engineering

Smash!

Tilt controlled wrecking ball toy

This interactive wrecking ball toy is controlled using a tilt-sensitive input via an infra-red communication link, allowing the player to swing the wrecking ball by changing the angle of the hand control. Developed using integrated circuit technology and programmed using C programming language, the toy provides a unique experience for the player, their gestures directly relating to the motion of the toy.

Mitch Gebbie
Product Design Engineering

Portamento

Synthesizer with an intuitive pitch bending interface

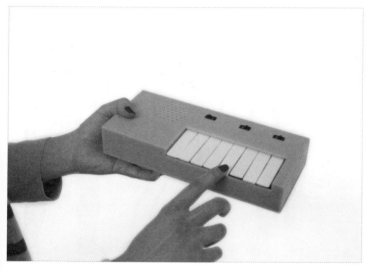

Portamento is a unique handheld electronic keyboard that has the ability to bend the pitch of notes. The user can change the angle of the keyboard with their wrist, which alters the pitch of the note being played. The change in pitch is controlled by a built-in accelerometer that detects the keyboard's movements and adjusts the note accordingly.

Andrew Davies

Product Design Engineering

Fox Camera

Exploring urban ecosystems

Inspired by the observation of semi-feral foxes in London this project aims to document urban animals through the creation of an autonomous remotely activated camera. Deployed in different locations it should unobtrusively detect the presence of urban wildlife using a camera and a passive infrared (PIR) sensor. Animal photos are safely stored in internal memory for later processing.

Dimitrios Stamatis

Industrial Design and Technology

Cowboys in Space

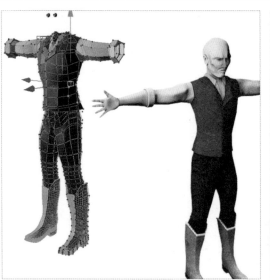

WEST results from a study of a wide range of advanced production techniques, and stylistic choices, which are quickly becoming more accessible to small independent creative groups. The media itself is a short trailer, for what could potentially be an independent feature-length film. Taking inspiration from larger budget productions, in terms of its cult film style and a whole host of cinematic techniques, WEST intends to entertain and captivate its audience through its bold style instead of high cost production. WEST takes advantage of full body motion capture technology, though a Viacom camera system, to record the actors movement in each scene. Other technology behind the trailer includes Xpresso script, poly-sculpting and sound design.

Xander Marritt

Multimedia Technology and Design

Mixtape Madness

Online UK music website

Mixtape Madness is the central hub for accessing a full archive of UK urban mixtapes. The website enables users to listen and download mixtapes for free, while offering artists feedback and reports on how consumers are interacting with their music. By using Mixtape Madness, artists will be able to track how users interact with their mixtapes and build an archive for new listeners to enjoy and enable them to download their whole catalogue. The website also incorporates video, blogs and an online community, reflecting the social side of the artists interaction with the listener.

Kingsley Okyere

Multimedia Technology and Design

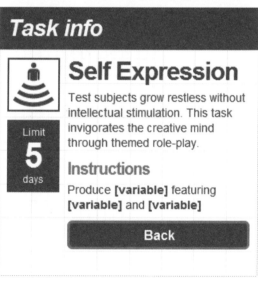

Pervasive games use reality as the game board. Players interact face-to-face in real locations according to a virtual layer of rules and narrative, commonly provided by an interactive, networked multimedia system. This project involved developing one of these systems, using the World Wide Web as a base. The ubiquity of internet access, especially on mobile devices, means a web-based approach has the potential to connect huge numbers of players from around the world, inexpensively, in real-time; a massively multiplayer portable experience.

Andy Green

Multimedia Technology and Design

Superclaymation!

Recreating claymation animation using CGI

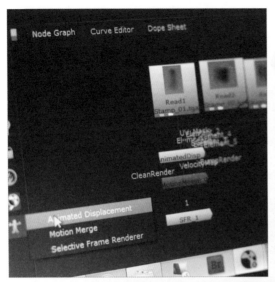

The traditional act of claymation animation, as used in such classic British animation as Wallace and Gromit and Tony Hart's Morph has been a process that has remained fairly unchanged since its inception. This project, involved designing a framework in order to attempt to emulate the look of this animation technique using industry standard 3D and 2D graphics packages.

Culminating in the production of a plug-in set for Nuke -a visual effects package developed by The Foundry- and a demonstration animation utilising full motion capture, the project raises important questions about the ongoing development and visual style of Computer Generated Animation as well as the place stop-motion clay animation takes in this changing climate.

David Paliwoda

Broadcast Media Design and Technology

Wacki Tap Tap
Captivating mobile gaming App

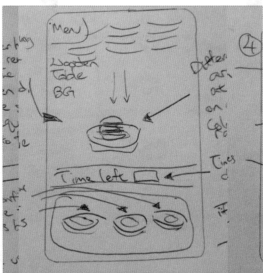

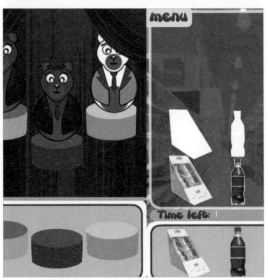

What kind of game features make a game captivating? How and why do people get addicted to gaming? The aim of this project is to create a mobile gaming App for iOS devices that contains various captivating features. The project was to create the App based on a combination of literature research (books, websites, journals, papers) and primary research exploring what features make a game captivating and how people get into playing games. The two sets of factors are then linked together to discover any connections between them. The results of this research will have direct input into the final App concept and implementation. The App is also a brand which can be further developed into other merchandise.

Jeff L K Lee
Multimedia Technology and Design

Thumbelina's Adventure

An animation inspired by the story by Hans Christian Andersen

This project is a graphic animation based on Thumbelina, a fairytale story by Hans Christian Andersen. This artefact exhibits character animation techniques, compositing, 3D camera movement, various screen transitions and expressions that are based on JavaScript language. The aesthetically expressive graphics enrich the animation. Such design will play with people's emotions, perception and imagination to ensure that the story catches the attention of the viewer who will have fun watching it. The animation style is influenced by the shadow-puppet style, an expressive art form of two dimensions.

Pavlina Blahova

Multimedia Technology and Design

Little Red Hood

An experimental psychological Flash project

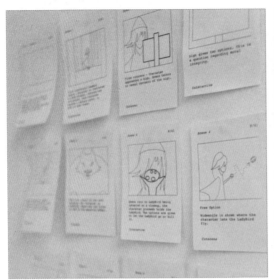

Little Red Hood is a cinematic Flash game based on the classic fairytale. It was originally a project to show off some animation skills, but it also became a great opportunity to explore interactivity. The game is dynamic, tracing the user's environment and immersing them in the world. The underlying feature is testing the emotions of the player and working out a psychological profile of the user in a simple format. The project was aimed to be used in psychiatry to help determine psychological profiles, however the project evolved into a concept, the first of a series of Flash games known as "Fairy Tale Adventures" that make use of basic psychology.

James Bruck

Multimedia Technology and Design

LITTLE RED HOOD

Why the Big Society needs Design

Evolving design to aid political agendas

Charles Eames, the American multidisciplinary designer was once asked 'What are the boundaries of design?' He responded: 'What are the boundaries of problems?'

The point that Eames was making is more pertinent now than ever before. The contexts in which design thinking is applied is increasing rapidly as society faces new and more pressing problems. The Design Council believes that although this does not appear to be like design in the familiar and tangible sense of the word; it could be the key to solving many of society's most complex problems and that there is a 'growing desire among designers, both young and old, to tackle society's pressing problems'. This may be just as well.

Relative poverty in Britain and social inequality have become critical issues with hundreds of thousands of families now trapped in dependency on benefits, facing considerable challenges to break free. Some argue that this is largely down to state welfare systems that worked to disincentive work and which restrict social mobility. The challenge is that these are not stand-alone problems. Social inequality and a lack of social mobility are considered to be direct drivers of mental health problems, educational failures, drug addictions, long-term poverty, family breakdown, long-term unemployment and anti-social behaviour.

How then does society go about addressing these issues? The Conservative Party offered one solution at the last election through something called the Big Society.

The core idea behind it is that, in the rapid growth of the welfare state, government policy has eroded the capacity of individuals, families and communities to look after themselves, but empowering people to deal with various problems locally will create a more cohesive and stronger society. Its supporters contend that in creating a state that does everything for individuals we have instilled the belief that others are to blame for difficulties and misfortune and that others will solve these problems.

The idea of empowering people is a noble one, most would agree. However, nearly two years after the introduction of the idea, only a small number of initiatives have been successfully implemented. The reasons behind this are complex, but one of the greatest challenges that David Cameron has encountered in trying to convey the idea is to explain the emotive reasoning underpinning it.

The Big Society requires local and personal interaction, so one general directive and policy cannot be fed down to every individual, family and community. Different people will have a range of concerns and reactions to the idea so for any real change to occur, this needs to be understood. This comes from understanding people on a personal and intimate level, their day-to-day lives, their worries and constraints.

This is an ideal context for design thinkers. Designers continuously work to understand a range of stakeholders and end users of products, systems and spaces. The interactions people have, the ways in which different people react and how they mobilise their environments vary incredibly, but designers must have the ability to understand what influences different reactions and how this can be controlled through design and through communication.

In communicating the ideas of the Big Society, designers need to understand the people affected by it, their expectations from a more empowered and local society and how to articulate these core ideas. Designers have the ability to empathise and relate to people, but also to communicate effectively based upon this understanding. It is these characteristics of design thinking and practice that may allow people to see past the 'politics' that colour perceptions of the Big Society.

Having the ability to understand those within local communities, relate to them and to help communicate core messages is a lovely sound bite, but if design thinking is to be applied to help put life into the Big Society, what skills do designers have for this? Furthermore, the application of design thinking needs to be considered practically in the context of Government policy.

Designers must be involved directly in communities to observe and experience all that individuals and families experience first-hand.

It is for this reason that it would be highly beneficial to place design thinkers alongside the team of 5,000 Community Organisers that has been created to help communities take control of services and systems in their local area.

Positioning designers right at the heart of communities would enable them to understand the needs and considerations of individuals. However, it is the way in which designers can use this knowledge to affect the way in which different people are mobilised by and interact with artefacts and environments that would enable the development of local initiatives.

Firstly, the approach that designers take to problems is different to that of policy makers, involving an open-ended enquiry that means no particular solution is actively pursued; instead, at the outset, the end result is not known. A design approach can move solving social problems away from being a process of decision making and policy direction to an iterative process based upon understanding users and changing the way in which they interact with one another and with social entities.

Harvard Professor Ronald Heifetz believes that "the most common leadership failure stems from attempting to apply technical solutions to adaptive challenges". A recent report by the RSA has used this argument as a focal point for explaining why the Big Society has failed thus far. The report argues that adaptive challenges require changes in attitudes and perspectives not just behaviours, and that the failing of the Big Society is that the public have viewed the idea as an attempt to solve socio-economic challenges with a technical solution. A design-led approach provides an alternative method.

Secondly, designers are proficient at visualising ideas and making them tangible. In trying to raise awareness of local initiatives and

developing a change in people's attitudes to social interaction, this design tool could be very powerful. Designers could be used to articulate the intentions of the Big Society and show it in action, but more importantly to articulate how it can affect individuals' everyday lives. The use of storyboards and communal discussion points are useful tools for designers not just in the way that users react to them, but also in how they can mobilise people and communicate ideals in an understandable way.

The success of the Big Society ultimately depends upon the levels of participation and interaction that each community can achieve with its members. Bill Torbet, a highly successful management consultant, once said "'If you are not part of the solution, you are part of the problem', is entirely misconceived... If you are not part of the problem, you cannot be part of the solution". This reflects, very accurately, a problem of the Big Society so far: if social interaction and responsibility are to change, people need more opportunities to share stories and have them heard, but also a supportive environment to have them challenged.

For the Big Society to become a reality in local communities, the emotive ideals and goals that are deeply built into the concept cannot be covered by the image of a 'top-down' self-serving policy. Currently, many local communities appear to still believe this is the extent of Cameron's message. Due to poor communications from proponents of the concept, I would have thought individuals, families and communities are yet to be well informed of what the Big Society is and what it is trying to instil within the UK

Patrick Bion
Product Design Engineering

FashionPad

A new and innovative fashion magazine experience

FashionPad is a new and innovative fashion magazine experience for iPad. You can download it anywhere, anytime: on a bus, at home or abroad. Traditional magazines are bulky and are limited by plain text and images that are often out of date by the time they hit the stores. FashionPad combines interactive video content, news that updates in real time and social media features, resulting in a fun and playful experience. Smaller overheads allow focus on the content. Topics can be more adventurous, offering different types of content from new, up-and-coming, designers to the business side of the fashion industry.

Johanna Elise Oja

Multimedia Technology and Design

SHOPSTAR

A time and money saving personal shopping assistant

Grocery shopping can be a tedious chore for many of us. With the current state of the economy, more people are cooking at home and being more careful with their money. SHOPSTAR is a mobile web application that has a total of nine features, which primarily aim to save you money and increase the efficiency of shopping at your local supermarket. Features include an item locator, basket total calculator, product comparison and access to past shopping receipts. This web application will give you a better idea of where your money is going and how you can tighten your belt … making your money go that bit further.

Mina Nishimura

Multimedia Technology and Design

Portfolio Personality

Aiding new designers in creating personal portfolio websites

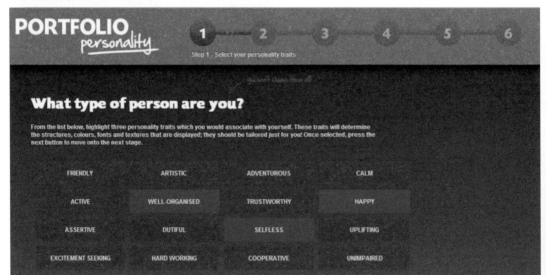

Portfolio Personality is an experimental project which uses an online step-by-step wizard to ascertain whether personality can be designed in portfolio websites, whilst also providing new designers a chance to create portfolio templates. Portfolio websites are arguably the best way for new designers to showcase not only their work, but also themselves as people. If it appears that people recognise certain personality traits within portfolio websites, then there is nothing stopping the notion of personality in design transgressing into websites of any genre.

Stuart Brockwell

Multimedia Technology and Design

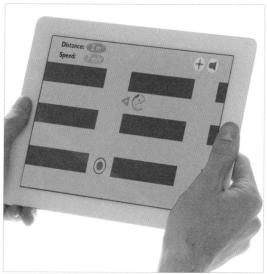

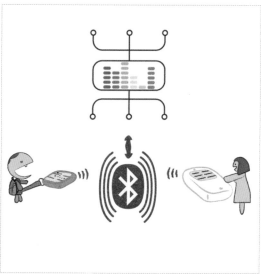

PacMom is an in-store activity for kids, which entertains and "lets them play" during the shopping process. It motivates them to stay close to their parents with the help of the augmented reality it creates. PacMom is a game like Pacman, in which a parent is tracked while they move around the store. The pathway is indicated on the child's device with dots and shapes. The goal for the child is to collect all the points by walking through them. As they achieve their goals, they follow the parent around the shop. The game also has different levels for different age groups that add challenges or tasks or even different games.

Ozgun Tandiroglu
Integrated Product Design

Modular Exhibition Stand

A new take on portable display stands

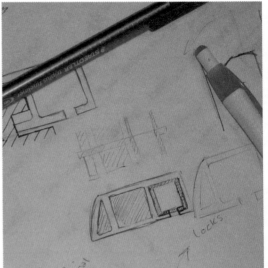

The use of portable display systems is common practice in many industries, with exhibitions being an important tool for selling or promoting new and existing products or services. Current systems, although providing excellent 2D graphic display, offer very little 3D display options and are limited in their versatility. This new take on portable display systems offers adaptable 2D and 3D display options, quick and simple construction and provides a level of versatility that existing systems fail to reach. The stand aims to provide the exhibitor with all that they could need, with integrated cable management, a fully movable platform, adaptable lighting, easily changeable graphics, and easy storage and transportation.

Sophie Richards

Product Design

Since the announcement that Rio de Janeiro will host the Olympics in 2016, the national and the local government have got together to make great investments in building new sports arenas and developing a new transport system (considered the best legacy for Rio) with new roads, routes and even bringing a new means of transportation for the city – the tram. However, the Olympic experience covers more than that because the customer starts his journey at the information touchpoints (official websites, hotels etc), going through the public transportation system and ending at the Olympic venues. Meanwhile, London is currently putting the finishing touches to host the 2012 Olympic Games and developing a project based on the Olympics during the current Games is a great opportunity.

Therefore, the project involves an investigation into how London deals with participants at the 2012 Olympics venues by means of transportation and way-finding signage. Having this as a basis, and considering practices adopted at big entertainment events, a complete, holistic journey experience strategy would be necessary for the upcoming Rio 2016 Games to reflect the brand values and experience.

The project intends to suggest a design-led holistic process for the Rio 2016 Organization (Local Olympic Committee and Local & National Government) to integrate the customer journey into the entire Olympic experience. To reach this outcome, first of all, the focus is placed on investigating the new transport system that is being created for 2016. In addition, a comparison with the current Olympic host city transportation system will be done. Moreover, an in-depth understanding of the Rio 2016 brand values and an examination of design experience practices at big entertainment events are necessary in order to reach the aim of this project.

Designing a meaningful customer experience, besides reducing misunderstandings during the trip to the venues, will rely on spreading the Rio 2016 values holistically throughout the journey, providing people a highly entertaining show and creating positive connections with the brand as well as strong memories for before and after the Olympic Games. Furthermore, the city of Rio de Janeiro is an exquisite location that will enhance the Olympic experience. The city image will also benefit from this, having a better image amongst locals and tourists.

Cilas do Nascimento Sousa

Design and Branding Strategy

Boonbox

The interactive, online gift finder

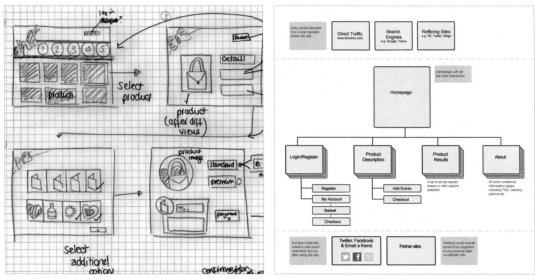

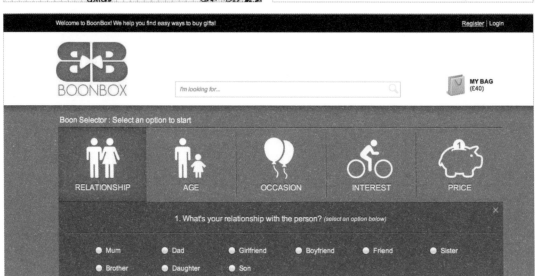

Boonbox is an interactive, online platform that allows people to find gifts for other people. It is often a hassle to find gifts for people if you do not have anything specific in mind, therefore we transformed the task of finding gifts online by developing an intuitive filtering interface that lets the customer build a profile of the person they are buying the gift for. Boonbox then searches a catalogue of products to find gifts that match this profile. Key focuses for this project were developing a strong visual language and a carefully crafted user journey.

L. Castiglione, A. Faber, M. Nishimura, J. Oja, K. Okyere, B. Williams

Multimedia Technology and Design

Online product reviews have just got personal. 70% of consumers read online product reviews before making a purchasing decision. The voice of the reviewer can have a significant impact on converting a visitor into a customer, but there is a growing concern with fake and irrelevant reviews. Releview aims to combat these issues and attempt to answer the two fundamental questions when reading a product review; 'How similar is this person to me?' and 'is this review genuine?' By incorporating a reputation and ranking system, we can sort reviews through the social relevance network based on similar preferences with the reviewer. Choices consist of filtering by experience, interests and reputation. Releview's mission is to make online purchasing decisions easier.

Balraj Chana
Multimedia Technology and Design

Smart Product Control Panel

Control panel for future smart products and the Internet of Things

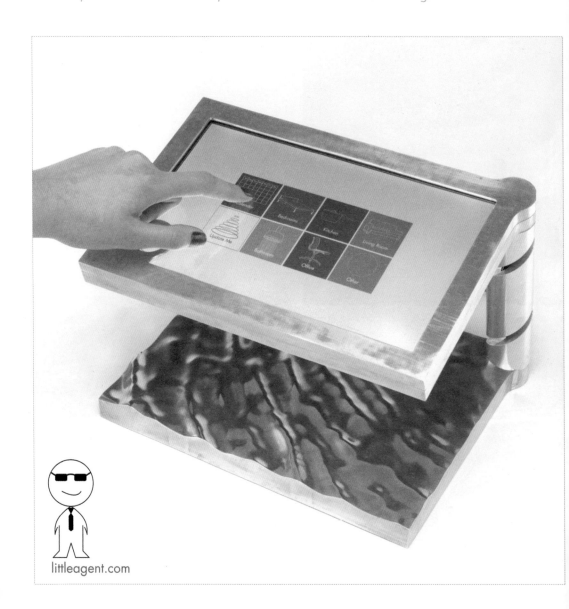

littleagent.com

Consumer electronics companies are developing new 'smart' products that aim to utilise the 'Internet of Things.' The problem with these new products is that they tend to consist of a display, control functions, and operating system, which differ significantly depending on the manufacturer. This inconsistency creates confusion, especially to those who find computers far from intuitive. This project focuses on the design of a control panel that will unify these 'smart' products by clearly displaying their information and settings. With the aid of hints and tips from the 'Little Agent' character, the control panel will easily monitor and control these 'smart' products using infographics, allowing interaction with a much simpler operating system.

Mark Taylor

Product Design Engineering

Site performance is a major concern in the Digital Industry. Studies have shown that responsive sites lead to a better customer experience and greater revenues for the site owner. Current industry tools to help developers monitor site performance require manual intervention and only provide a snapshot of performance. Marlin is a revolutionary application, which is fully automated. It provides detailed stats, allowing the developer to see clearly whether site performance is improving or degrading. No other application gives this critical insight during the development phase. Developers save time. Companies save money. And users enjoy a less frustrating web!

Aaron Faber

Multimedia Technology and Design

Low-budget supermarkets are becoming increasingly popular in the UK retail grocery industry. Although some of these brands' offers have tempted away a certain amount of budget-conscious customers. Customers of premium supermarkets still tend to be more loyal due to the quality of the shopping experience they receive. The lack of loyalty has become the major problem that is restricting the growth and development of low-budget supermarkets in the UK. Therefore, how could we draw back customer attention to low-budget supermarkets and build customer loyalty through experience design?

This research aims to suggest a design-led brand strategy for low-budget supermarkets in the UK through the use of experiential design in order to build a long-term relationship between the customer and the brand and enhance customer loyalty. The objectives of the project are meant to guide into the research process. Therefore, the first step to be taken involves the identification of current and potential problems and challenges that budget supermarkets are facing. Furthermore, the research will look into experience design, customer loyalty and customer behaviours for a better understanding of the concepts, followed by investigating the concepts of brand awareness and

brand experiences and the factors that influence them. Moreover, identifying how environment, services and ease of shopping may influence brand experience are essential as well as analysing examples of best practice.

Finally, a proposal of effective branding strategy will be made, meant to engage customers through a holistic design-led experience and enhance customer loyalty.

> *The overall project result will provide economic and social benefits to low budget supermarkets.*

The research will focus on low-budget supermarkets in the UK, such as Aldi, Iceland and Lidl. Both primary and secondary research will be carried out to obtain reliable information. Methods of research will include interviews with professionals, in-store observations, survey and questionnaires, customer review, and case studies. Beneficiaries include employees, customers and budget supermarkets, providing them future development and growth opportunities.

Wing Shan Michelle Poon
Design and Branding Strategy

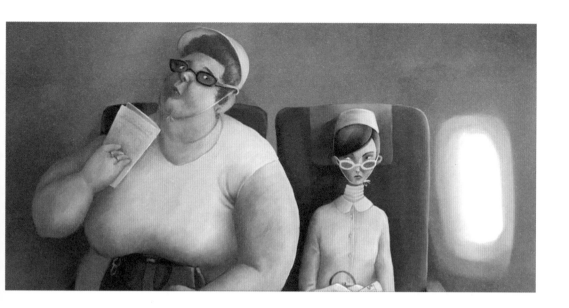

Obese passengers have difficulty fitting into a single economy class airplane seat. This is uncomfortable for them and other passengers around them.

> *In 2009, 23% of the adult population in the UK was obese and experts believe that by 2030, this number will rise to 45%.*

The World Health Organisation describes this global epidemic of overweight and obesity as 'globesity'. Some airlines have implemented a policy for bigger passengers; others have not, or are in the process taking this unavoidable step. Unfortunately, even the airlines that do have policies on accommodating 'customers of size' on their flights, do not always communicate this information to their passengers in the best possible way.

Research shows that issues arise at all stages of the flight experience; from arrival at the airport until leaving the airport upon arrival at the destination. Various airline websites do not provide information about seat dimensions, neither do they support the option of booking 2 economy class tickets online. Bigger passengers also experience greater difficulty when walking long distances to gates, sitting in the relatively small seats in waiting areas and walking through the aisle of the aircraft to reach their seat. Once seated, there are other issues such as lowering the armrests and the tray table, making it difficult to eat or drink aboard the plane. Also, having to ask the cabin crew for a seatbelt extender can be humiliating, an experience intensified by the fact that some airlines provide brightly coloured seatbelt extenders. All of the issues mentioned above can be resolved by applying better product design, service design and experience design. As the number of obese people is increasing, there is no way to avoid changes being made to airline policies and products. Short term solutions should involve changes to the airline website and policy. Where long term solutions should include physical changes to accommodate the changing demographics, such as alterations to airplane aisles, seats and tray tables, as well as airport seats and transport tasks.

This project is an investigation into the aforementioned issues and suggests viable solutions to make the flight experience of the larger passenger more comfortable and enjoyable.

Floor Veldhuis

Design Strategy and Innovation

I <3 You - Love Online

The impact of the Internet on the pursuit of love

We are all aware of the increasingly large part that technology is playing in our lives. The digital age has affected almost everything about us, including that most basic of human desires – the desire for love. Our search for love is as strong as ever – match.com found that nearly half of UK daters still rate it their number one life priority – but our approach to love and relationships has altered beyond all recognition. Communication is vital to all relationships, so, as the Internet creates new ways of communicating, it has a direct impact on how we relate to one another in the pursuit of love.

Many feel that the age of the Internet and smartphones has killed off traditional romance. The era of the handwritten love letter has indeed largely gone, but is it true that romance is dead? Throughout history, communication channels for love and romance have changed as new technology is introduced. Morse code, carrier pigeons, newspapers, couriers, the postal service, telephones, fax, email, you name it; couples utilise them all in the exchange of romantic affections. Online communication is simply speeding up the traditional method of writing, closing the time gap between sending and receiving. We no longer have to wait weeks for a letter in the post: conversations can be instantaneous. In this new era of flexibility and choice, has the Internet made the pursuit of love any different?

One of the most popular uses of the Internet is in the search for relationships. Internet dating has grown significantly, attracting a much larger user base and offering increasingly sophisticated self-presentation options.

Research carried out by match.com, found that singletons in London were pickier and had higher expectations, possibly because London has more single people and therefore more choice. Online dating has the ability to replicate the large dating pool found in big cities, and can reach even further afield, offering the opportunity to find a 'soulmate' on the other side of the world. Now, as the popularity and participation on these sites grows, the stigma of using them decreases, with online dating now the third most popular way to meet a partner in the UK.

Many criticise these sites, pointing to the loss of face-to-face interaction, but in this respect, it is not dissimilar to the days of letter writing and other text-reliant communication. During any communication, when the parties are not physically together, they will use their imagination to fill in the gap normally fulfilled by vision.

A conversation may appear to be going brilliantly, but is it just answering to our delusional expectations?

As a result, research has found that people are more comfortable giving compliments in text communication, saying 'I love you'. People feel safer offering more information and discussing personal topics: 'So how did your last relationship end?' Long online relationships like this can leave online communication at risk of the boom or bust phenomenon. The increased self-disclosure moves the relationship on quite quickly, with heightened feelings of intimacy, further fuelled by imagination. The connection can feel strong, but upon meeting, reality proves not quite the same.

The anonymous nature of text-based online communication can appear very attractive: no awkward silences, no blushing, no painful rejection. Alone in a room, with just the computer screen, the attention is focused entirely on the writing. All of the communicative energy that would be used in watching the person, other people in the room, the environment you are in, is redirected to the message being written. When a relationship is initiated online, everyone has the same voice and is judged by the same standard: their words. People find great comfort in this as it can remove the hostility of the more personal side of natural reactions.

Approaching members of the opposite sex is a notoriously nerve-wracking experience. It is no wonder that the ease in using text-based communication is preferred – but it is this that presents the main difference to real life conversations. On a date in the real world, the amount of time a couple spend talking is surprisingly short, considerably less than in the word-based cyberworld. This is not all down to awkward silences... most of our time is spent reading and understanding body language, facial expressions and other non-verbal cues. Our capability to read reactions, to detect subtle changes in a tone of voice or a flick of the hair, is innate. We have evolved to watch and decode the behaviour of others, whilst the reading and writing skills of text-based communication take years to master. In online communication, some senses are lost entirely. We have the technology to live stream video, listen to voices over the telephone, but what about smell, touch, and even taste? These are our most primitive, but often strongest, senses whereby relationships progress to deeper intimacy. Live video chat seems the most successful form of online communication in mimicking real life. You can see facial expressions and body language, and hold synchronous discussion. But with awkward camera placement, you cannot look into each others eyes, and the power of touch and smell is still missing.

> *"My favourite thing about the Internet is that you get to go into the private world of real creeps without having to smell them."*
>
> Penn Jillett

At least the aim of dating websites is to eventually move the relationship offline. Because of this anticipation of face-to-face meeting, communication here is considerably more honest than in chat rooms and other websites. Text communication proves useful in judging whether you have similar interests and values: a highly rated part of a relationship. However, the most important factor for a successful relationship, as match.com found rated by UK daters, is a feeling of security and sexual compatibility, which require a meeting in person. Joseph Walther, an expert in computer-mediated conversation research, advises that if a strong contact is made, move it offline as quickly as possible to prevent idealisation. This will reduce the possibility of disappointment and hopefully romance and chemistry will follow offline. With the ability to match people with others they may never have had the chance to meet in reality, online dating serves as a tool for making connections. With busier lifestyles and less time, people of all ages are finding it increasingly difficult to find time for their personal lives. The Internet provides a versatile platform that opens up a world of possibilities, allowing you to cast your net much further afield.

The need for love and relationships is part of human nature and will never change. It is just the technology we use to get there that is constantly evolving. We will continue to use anything that helps us achieve our goal, which in the end, is two people, alone together, in love. The Internet has not killed romance, it is simply a tool to fuel it.

Sophie O'Kelly

Product Design

Activity Booking System
A new method of communication within public spaces

People now have more choices of how to communicate with each other but a lack of real face to face interaction is becoming a serious issue. This project is an investigation looking at providing an opportunity for people of different generations to make new friends and communicate with each other in social public areas. This project is a booking system, located in kiosks in public parks, that enables people to identify and register with activities and social events happening in the park. People can check what events are taking place at the kiosk and book events which they are interested in. The system affords a range of different activities including social and sporting events for the public.

Xuefei Zheng

Integrated Product Design

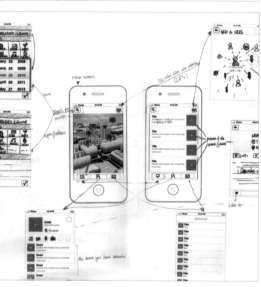

This iPhone application provide people with a way to show and share their interests. It aims to help people find someone who has similar interests and to allow them to meet each other in public places. BuddyOn has three main functions. The first is to search the events in your local city. These events are sorted in different interest areas such as music, sports and exhibitions.

The second is to see other users around you providing a way to find a buddy who has the same interests as you. The third function is mainly designed for business users, for example, the owners of bars, book stores or galleries. They can check what local people have liked in recent times and improve their business accordingly.

Chao Luo
Integrated Product Design

Interactive Wallpaper

Bridging interior design with fun and creativity

In the house we are often strict with children, making sure they are careful not to paint or draw on the walls. This concept of interactive wall paper is about the exact opposite. This provides the opportunity for children to express their creativity without restrictions and boundaries but also bridges the gap between them and adults with a more fun way of communicating.

This wallpaper can be reformed through interaction between users and the space. In this way, the fear factor will not only be eliminated but also the space will eventually transform into a personal creation, far more inspiring than any readymade pattern bought through conventional wallpaper products.

Polina Liarostathi

Integrated Product Design

Gong

The public bench for the better communication in airports

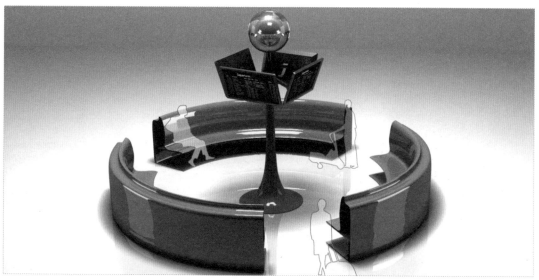

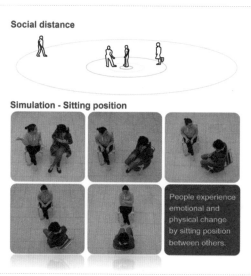

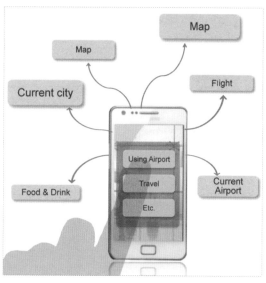

Lack of communication in public areas has become a serious social problem with the development of technology and urbanisation. There has been a particular increase in airport users in recent times: 69.4 million passengers arrived at Heathrow Airport in 2011. It is understandable that there are many problems relating to lack of communication, information and places to rest. The aim of this project is to make the airport experience more efficient and convenient through the use of public furniture and smart-device interfaces. The bench is designed based on the social distance and sitting-position simulation, providing the suitable environment for people to feel more comfortable with others. Smart-devices can share and get the information from the airport and other users.

Yong Lim

Integrated Product Design

Design your city

Implementing street art inside the urban environment

Nowadays, society witnesses the increasing popularity of street art, mainly as a consequence of commercialised and over controlled urban living. Street art is a unique modern counter-cultural form of engagement with the urban realm but it is still mainly seen by the authorities as an aggressive act on the established urban system. Even though constructions and corporate billboards arguably disrespect public space more than the street artists, the question is still very widely discussed by the cities around the world:

'Who owns public space?', and how can a country be a true democracy without public space being actually public?

The street art trend is what happens around the world, with many exhibitions, festivals, and even whole districts dedicated to this form of art. Even though, some city authorities have already realised the potential of legitimising street art, the advantages of these public interventions are still not widely exploited. Street artists speak the language of the streets and have passion to improve dull city landscapes, but the main advantage of this form of art is an element of play and interaction between artists and citizens, extending the reach of art to a wider and more diverse audience.

London`s authorities have already implemented bottom-up governance, actively involving citizens in designing the city landscape. Therefore, taking the open-minded nature of the city`s authorities and public appreciation for street art, this research is focused on enhancing the possibilities of such a controversial form of art and the extent to which it can be applied to all manner of city regenerations and branding; but most importantly it proposes ideas on how the main target groups (local authorities, property owners, street artists, and citizens) can collaborate to develop the urban regeneration agenda around concepts of social inclusion and street art.

Overall, this research provides social and economic benefits to all stakeholders. The co-design platform will give voice to the citizens and improve the city brand policy that can help to actively engage citizens in designing their own environment, helping to create out of abandoned areas attractive ones and endorse London`s creative ambience.

Yana Zalesskaya

Design and Branding Strategy

St. Etienne Multi-Disciplinary Project

Building concepts for the future in a multicultural environment

Following a four year old tradition, the École Nationale Supérieure des Mines de St-Étienne in collaboration with Brunel University London brought together design, arts and engineering students for a week of multidisciplinary workshop. The 2012 co-creation process ran under the label of "smart, adaptable, functional textiles" and encompassed an exciting journey from brainstorming, to research and ideation, to presenting the concept, and finally to building prototypes and/or business models for a smarter future. Amongst this year's innovative ideas have been the wall of sound, the high-tech food packaging textile, the illuminating grass fabric, the smart shopping bag and many more.

Design Process

Every Brunel design thinker has their own way of working, but there are common themes across all of our design processes that can be explained through the double diamond model. This structure is outlined below in the following set of images, depicting the kind of questions we were asking ourselves throughout our projects.

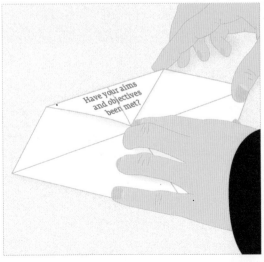

Future Urban Luxury Hotel Room

2025 scenario for a luxury hospitality experience

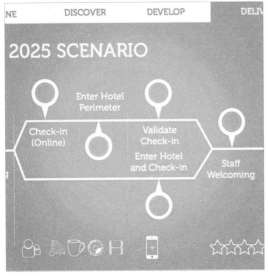

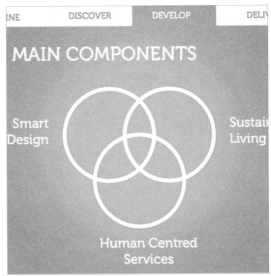

The entire concept of luxury has been redefined over the recent years and customer demands became more complex. In the past, the concept of luxury in a hotel room was product centred, technologically limited, with main focus on a classic style through ornamentation. Todays room offers more than that. It all revolves around the customer and delivering the most exciting experience, adapting to his/her needs, offering options and customised services, a simple room design, all in a human centred approach.

The research started from defining the main mega trends that are and will be influencing the hotel industry: connectedness, indulgence, ethical consumerism, high expectations and individuality, feminisation and tribalism. Based on these aspects, further research and discussions with experts have been conducted resulting in defining three key components that will drive the urban luxury hotel room experience in the future.

Firstly, the entire experience and design will be human centred. As the current context has become more permissive for people to travel, their expectations from a luxury hotel increased. However, they do not seek complicated features, but instead simplicity. As Karen Rosenkranz, Head of Social & Lifestyle Foresight at Seymourpowell mentioned, people "especially frequent travellers, don't want endless choices for everything". But instead they will care more about details, inclusivity through design and personalised service. People will have an input into what they should be offered. Secondly, through smart design, space will be more efficiently used (modularity and adaptability) and technology will facilitate the customer experience, engaging people through virtual environments. Thirdly, sustainability will be a core value of the hotel. Aiming to bring a local flavour of the city/country, hotels will be working closely with local craftsmen and materials, reducing waste and protecting the environment.

Based on these three components, a scenario for the year of 2025 has been developed. Thus, the entire journey will start from the room booking moment when the customer can create his own personalised profile that will stand as a basis for the room experience. By using his smartphone he will be able to activate various features inside the room such as chromatics, sounds, temperature. Motion sensors will enable having gaming and gym areas in the room as the walls and mirrors will be displays to connect with virtual environments, act as TV and computers at the same time. Everything to pamper and intimately connect with the guests.

S. Chen, C. Montenegro, A. Pîrvu, C. Sousa, J.E. Su, V. Tseperka

Design Masters

Tony Hsieh, Zappos's CEO believes that "If you get the culture right, most of the other stuff, like building and enduring brand, will happen naturally" while Naomi Stanford contends that "How much the purpose and values of an organization matter depends on how well they are experienced by it's stakeholders."

Powerful and accessible information communication technology has transformed our world in an irreversible way. The Internet and mobile developed a network where each connected individual has the power to send and receive messages making it impossible to manipulate content and media. In an environment like this, transparency is not an optional value, but a scenario feature and a basic survival competence. We are living in the Information and Knowledge Era where "brands have to walk the talk" . At the same time, Brazil is living a moment of fast economic growth. Besides new challenges such as mergers and acquisitions, companies need also to learn how to survive and differentiate in this new scenario where consumers are being empowered through information transparency and sharing. As a consequence design and branding consultancies started offering new design solutions in order to redefine brand strategies, strengthen company identity and deliver real value to customers. On the other hand, no matter how good the brand strategy is, if the organisational culture does not support it, the strategy will not be truly implemented as 'culture eats strategy for breakfast" according to Peter Drucker. Understanding and managing the corporate culture in order to align or even transform it became a highly important skill, a source of competitive advantage where strategic design plays an important role. However, the knowledge and investment in strategic design management within Brazilian companies is very limited. As a consequence, many branding projects are only developing tangible assets such as visual identity, but the corporate culture does not reflect yet the brand values.

This design research aims to answer the question "How to apply strategic design in order to make the corporate culture reflect the brand values?" The expected outcome is a design-led tool for branding consultancies to apply within Brazilian companies in order to facilitate corporate employees' understanding of branding and enhance brand culture. Increasing the understanding of branding within companies, facilitating the branding process and enhancing brand culture are the main benefits of this research project. The beneficiaries are branding consultancies and Brazilian companies going through a rebranding process.

Carolina Montenegro
Design and Branding Strategy

Branding Bran

A catalyst for enhancing tourist experience

The significance of place branding and the positive impacts it has upon a country's capital and indeed for tourism is becoming a much debated field of study. We are living in a globalised world where every country, region, city and even village has to compete for its reputation, its share of attention, goodwill and trust. The academic expert whose publications served as a guide in conducting the research for Bran Village is Simon Anholt, inventor of the "nation branding concept". The quote which inspired the overall research paper was given by the same branding expert:

> *"Place branding in its advanced form is primarily about people, purpose and reputation."*
>
> Simon Anholt

Throughout the research that has been conducted for this project it has been identified that in order to be competitive in the global tourism marketplace there is a need to provide compelling tourist experiences.

Bran village, a place of myth and legends - the place where the story of Count Dracula was born - situated in a spectacular medieval landscape, is currently experiencing an unexpected growth of tourists, all demanding unique experiences that only an iconic place like Bran could offer. Count Dracula's Castle, the outstanding landscape, the unspoiled nature as well as well-preserved customs and traditions which are part of the beauty of the village, cannot be fully experienced by tourists due to the lack of engaging activities. Therefore, the problem identified in this case, is that Bran village has a brand strategy which is unclear and confuses the tourists. Also, the lack of international standard tourist services and facilities affect the overall quality of tourists experience. The solution to the problem is to create an effective design-led brand strategy to act as a catalyst for enhancing tourist experience. Thus, the question that needs to be answered is: how design and brand strategy can satisfy even the most sophisticated consumers?

The research methodology focuses on qualitative methods: ethnography, focus group, interviews, personas and scenarios. These all are meant to better understand the consumer: whether it shows the desire to know more about Vlad the Impaler aka Count Dracula and his role in history or whether the consumer is an architecture design lover or a going back to nature nostalgic. Understanding the people, having a clear purpose and building on reputation will lead to an effective design-led branding strategy for Bran village.

Georgiana Bădălîcă-Petrescu

Design and Branding Strategy

By 2020, the world population aged over 65 will have trebled from the present number to 700 million people. Based on the Ageing Population Trend and the following significant mega trends which have been identified during the research such as: 'Globalisation and Localisation' and 'Advanced Technology', the project aimed to create a Cosmetic Medical Tourism platform through design for 2050. The methodology framework consists of both primary and secondary research. The literature review is composed of theoretical and practical data and serves as secondary research, whereas interviews and the Delphi method represent the primary methods.

Through the research, it has been discovered that the ageing population is placing high emphasis on youthfulness which as a result is arguably making the cosmetic surgery the fastest growing component of the healthcare system. Undoubtedly, the plastic surgery industry is developing, facing a shift in industry locations from beauty centres in the USA to those in countries in Eastern Europe, South America and South Asia, where high quality surgeries at lower prices are offered. Based on key findings, "the holistic package tour in 2050" scenario has been created consisting of three parts: before, during

and after surgery. For the first stage, the "Forever Young Application" was designed to be used via several devices such as Tablet PC, Smartphone, as a service for consumer to research and compare the medical surgery tourism. Furthermore, in the second stage, the development in techniques and advanced technologies such as stem cell, bio-printing, holographic, robotic will be embedded in the plastic surgery industry in the future. In the last stage, there will be an increasing focus on spirit recovery. The hospitals will provide medical service and amenities similar to luxury hotels. Furthermore, plastic surgery and travel will be combined and be introduced as a cruise ship care.

In conclusion, Medical Tourism, especially referring to the plastic surgery niche will become a big market within the next 50 years. The emerging technology will revolutionise surgery and communications techniques. Moreover, the increasing demand on spiritual experience will drive a change in the service system. Thus, the country or organisation that can provide the holistic medical tourism services to make the journey convenient, efficient and enjoyable, will gain advantages in the highly competitive market place. Design will play an integrative role delivering communication, experience and user well-being.

G. Bădălîcă-Petrescu, K. Amnueypol, M. Jang, S. Chantathab, Q. Chang, N. Chang, Y. Tien
Design Masters

Building emotional connections with cosmetics brands

Brands have overwhelmed contemporary society. The power of brands is undoubtedly great since they can formulate ideas, attitudes and even whole subcultures. They spread in different industries across different fields of products and services. They tend to differentiate themselves through their various component elements and with which people interact. According to Wally Ollins, a brand's tangible aspects consist of four main components: product, environment, communication and behaviour. These main components play a key role in the outcome of the project as they are used as the core of the framework to be developed. Building strong brands and emotional relationships through holistic experiences seems to be the most effective way of differentiation especially for brands which are connected with self-esteem, personal fulfilment and the personal indulgences of the customers such as cosmetics brands.

The cosmetics industry is a highly competitive environment based more on fulfilling emotional needs through its products, especially for female customers. Living in a society that has the ability to provide a wide variety of cosmetics, especially women-oriented beautifying cosmetics, customers' perception of brand seems to be an essential element that contributes to its success. However, cosmetic brands seem to neglect this perspective and, as a result, customers can't really distinguish between brands and get really attached to any specific brand. According to researchers, teenagers are an upcoming market segment. Female teenagers tend to spend a lot of money on cosmetics products given that they try, through them, to cover some of their emotional needs such as boosting their self-esteem. Furthermore, different studies support the fact that at this age people build their brand loyalty. This is the reason why this project is focused on teenage girls.

This research examines the different theories that can support emotional connections and holistic experiences between brand and customer through an extensive literature review. The current practices used in cosmetics branding were investigated through case studies of brands that are beloved. Also, the project tries to identify customers' requirements and key factors influencing customers' brand perception and emotional connections through primary research consisting of interviews with youth experts, questionnaires and focus groups with teenagers. Last but not least, Wally Olins's four vectors framework was chosen and adapted in order to be used as a basis for the development of a design oriented brand framework that can enhance emotional connections with customers.

Vasiliki Tseperka

Design and Branding Strategy

Have you ever bought the Big Issue? Or watched Jamie Oliver's Fifteen on the television? Then you already know what a social enterprise is: a business that trades as a way of to improve communities and people's lives, to tackle social and environmental problems. However, in their pursuit to determine change and innovation most social enterprises do not have a strong enough brand image that would engage and motivate people to participate in the process.

> *"Good design process focuses on the inter-relationship between users, workers, professionals and services"*
>
> *Design Council*

Co-creation is an emerging tool for engaging people in building meaningful services and products and creating shared value. As Tim Brown, CEO and President at IDEO mentions, "design has become too important to be left in the hands of designers". Thus it is important to bring together stakeholders – business partners, employees, citizens and others - in designing solutions collectively and locally for various issues. At the same time, co-creation not only builds service experiences but can re-energise brands, being intimately connected with the

brand as Nicholas Ind affirms. Therefore, if people who will use the brand create something meaningful for themselves they will bond even more with the entire brand ideology. And the ideology is the one that sets the framework for innovation. The research intends to look into the extent to which co-creation can contribute to building strong social enterprise brands and design better service experiences. The approach involves several levels: emotionally, functionally and socially.

The research methodology is mainly focused on qualitative methods through which the attitudes people have towards co-creation are being investigated as well as their perception of social enterprise brands. Findings so far revealed that people are willing to engage in co-designing for social causes and online tools such as open source and crowdsourcing platforms seem to facilitate this best. As Dr. Mike Short mentioned during his visit at Brunel, "the future belongs to the open source" and these tools make it easier for people to create strong, meaningful social enterprise models that encourage user involvement. The expected outcome of this research is a design-led strategic framework for translating the value of the co-creation process into meaningful brands for social enterprises, empowering people to become designers of their own solutions.

Alina Pîrvu
Design and Branding Strategy

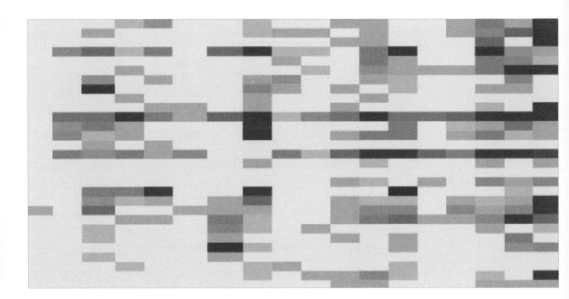

People are becoming increasingly interested in what they are endorsing as a consumer. What is their money doing? Where does it go? Is buying power being used for good?

Corporate Social Responsibility is the term which is given to the way a business conducts itself in relation to the expectations and social norms of society. To be responsible in the way it conducts itself and to be sustainable within the world of business and the outside environments as well. The case for CSR in business is often fought by trying to prove that it increases profit so it is always in the best interest of business to be responsible and ethical and sustainable.

This approach has proved to be sound, but it could be suggested that we as a society and our collective consciousness are becoming much greater as the years go on so that we now think that we have a social duty to life on this planet. This duty cannot be ignored by anyone and if we are to keep up our high standards then sacrifices must be made.

Is it not shameful that we put prices on the health and lives of others in other countries? Trying to change top down is not going to work, we can applaud people trying to change but in reality it is young companies which start with the right principles, ideas, ethics and sustainable practices which will be the great companies of the future.

We consume in an ignorant haze

The aim of this research is to establish an industry wide understanding and definition of what Corporate Social Responsibility currently means. Once established, a comparison of this to early definitions will highlight the progression and analyse what has made the definition evolve. The definition will be part of a larger research into how to build a company's Brand Equity on Responsible, Sustainable and Ethical principles. The research will particularly focus on these developments and initiatives in the UK's fashion industry.

Benjamin Buist
Design and Branding Strategy

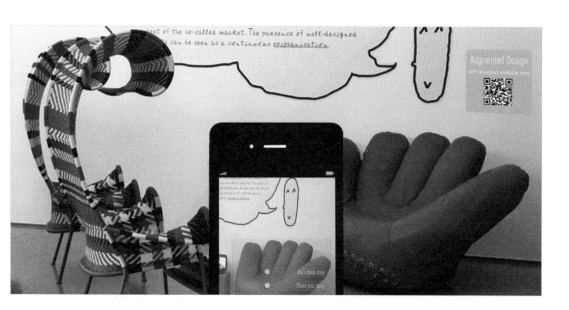

Experts agree on the fact that the industrial economy model based on manufacturing that dominated the twentieth century is worldwide giving way to the creative and knowledge-based economy model which places high value on human creativity and gives prominence to innovative ideas above all else. It has been proved that cultural and socio-economic development are not disconnected and extraneous phenomena: they are part of a larger process of sustainable expansion. In this context it is fundamental to foster creativity and innovation through a set of consistent actions. In particular, after the economic recession, there is a widespread need to invest in arts, culture and creativity in order to prevent the loss of an entire generation of creative talents.

Museums have always been repositories of past creative expression but for a long time they have also been considered as authoritarian and severe institutions. Back in 1975 Jan Jelinek, museologist, wrote: "Museums only fully develop their potential for action when they are actually involved in the major problems of contemporary society". Therefore, if museums want to stay relevant in contemporary society, they will have to evolve to meet society's need for creative incubators for young creatives.

Design museums have the opportunity to achieve their full potential as exchange and innovation platforms encouraging individual creativity and therefore contributing to a sustainable development in contemporary society. They can do this providing inspiration, but also tools that are typical of the design discipline. As Eleonora Lupo, researcher at Politecnico di Milano, stated during an interview: "Design can easily provide the 'content' to design museums, but it should transform them in 'contexts' of innovation as well". Strategic design will have a key role in this process of evolution as it can help cultural institutions envisioning opportunities and intuitions and designing integrated solutions.

The purpose of this research project is to develop strategic guidelines for European design museums in order to make the most of their potential as innovation platforms and to investigate at what levels strategic design can drive this process. The project aims at identifying the main issues and design-driven solutions and at offering economic and social value envisioning a future development opportunity. To some extent, both the design museums and society will benefit from the research, but further studies will be required to generalise the findings and ensure their transferability.

Sofia Carobbio
Design and Branding Strategy

Breadth, Depth and T-shaped Designers
What should we aspire to?

One debate that rages on between design professionals from various disciplines is the definition of a specialist, polymath and everything else in between. Should a designer be a specialist in his or her field? Or should they have knowledge and skills that span across various areas of design and beyond.

The term 'T-shaped' originated in America from IDEO's Tim Brown. In his blog article named Strategy by Design he describes T-shaped designers as having "a principal skill that describes the vertical leg of the T. But they are so empathetic that they can branch out into other skills, such as anthropology, and do them as well.

This produces a designer who surely has a better grasp on the expansive landscape of modern design. Or, within this destination, does it produce a designer with such a broad scope of skills and knowledge that they become a jack of all trades, and a master of none. These interests and skills could be directly design related or they could be related to broader topics, for examples travel or sport. To quote Richard Hartle, an accomplished 2D and 3D designer, "design is often about exploration, curiosity and experimentation so I'd say it's extremely unlikely that you'll find a designer who doesn't have other keen interests , even if they are 'arts' based." He goes on to argue that the human condition is to be curious, so surely all designers will have interests that span across different domains and that by doing this it will improve the designer's ability to design.

Another argument is that having a broad range of design skills can help one to get a better grasp on a project than a specialist could. Richard Hartle is of this opinion expressing "Having some knowledge of a range of design areas means I have a starting point for research and exploration rather than starting at a blank page. Also being skilled in a number of aspects of the design industry such as furtiture, interiors and graphic design allows me greater access to interesting projects." However if a designer is happy in their specialism who is to say they should broaden their skill set? If learning is more effective when you enjoy the topic, then is there much point in forcing yourself to learn a new skill set that you don't enjoy?

The debate about what makes the perfect designer is perhaps especially significant to a design graduate. Of course every company is looking for slightly different things, but what type of designer should graduates aim to become? "These days, as the design industry gets more competitive, we just cannot afford to come up with similar design solutions." This opinion from Brian Ling suggests that two designers with similar skill sets would both produce similar outcomes, and therefore a designer with a broader range of interests and skills could be more attractive to an employer.

Another way of looking at this debate is with the question, 'what makes you original?' It could be argued that a graphic designer with knowledge of product design can only have an advantage over a designer with merely one skill set. Having a more unusual mixture of knowledge and skills should create a uniqueness that can only create more interest from employers because it makes you different to any other designer. Passion is an important element of design. A designer with no passion for what they do can never be as accomplished as one who sees design as more than just a job. Richard Hartle sums this up well by saying;

"It depends on how you see design - is it a 'job' or a 'way of life'?"

So to become more T-shaped, it is important to work with people who's core skills lie in other areas. By working along side professionals from areas other than your own, it is possible to gain an empathy for their trade and also learn some of the skill sets involved. This is expressed nicely by Glenn Angelo, a designer at Essence, when he says, "I try to surround myself with intelligent, creative people, no matter their profession - because there's always something you can learn. Remember, if you're the smartest person in the room, you're in the wrong room."

Leonardo Da Vinci could be described as a polymath because of the amount of things he excelled at. As Tim Brown puts it, "as the design problems we are being asked to tackle get ever more complex, we can no longer operate under the false assumption that we as individual designers can craft appropriate solutions. Rather than aspiring to be polymaths or modern day Leonardo's, it is being T-shaped that is the key to success in the expansive landscape of modern design." Although it seems the T-shaped designer is the ultimate balance, there have been several developments to this model. Instead of aspiring towards being T-shaped, another alternative is to become m-shaped. Jack Moffett, an experienced Interaction designer believes that "a mature designer should begin branching additional verticals, moving from a T to shapes more resembling the letter m." For a designer to remain relevant in the industry, he should continually seek new areas in which to add value. Kevin McCullagh sums this up well when he writes, "Serious designers should think twice about playing up their horizontal skills, and instead get down to the tougher but ultimately more rewarding work of consciously defining and building the new verticals for emergent design disciplines."

What makes you original?

So the debate rages on. Specialist, generalist, T-shaped or m-shaped, what is the perfect blend of designer? It seems that the most important thing is to love what you do, keep learning and opening yourself up to new areas and information and surround yourself with talented and inspirational people. But most importantly, show what makes you original because everyone has a unique blend of talent, skills and experience to offer.

Jesse Williams
Industrial Design and Technology

Each year, we carefully select a brand as a starting point for exploring the world around us and what this might mean for the future, what will be relevant and important for the brand. We refer to this work as 'Contextual Design'. The results are grouped here under the title of Brand Futures. Each team's project results in a series of possible future products for the selected brands, or what can also be termed 'brand extensions'. An informed and intelligent prediction of what these companies would be doing in the future.

We demonstrate how product and industrial design can be an incredibly powerful tool for bringing future visions to life.

For each brand the designers have explored and gained a deep understanding of the brand's DNA: the irreducible qualities that should be traced in everything the brand does.

A thorough understanding of the brand is mixed with structured investigations into the macro environmental factors which will be impacting on the brands over the next five to twenty years. Ideas are then developed for a refined brand positioning taking these factors into account.

For example the Fred Perry brand team found a match between the strong heritage and integrity of the brand and the need for a shift away from 'fast fashion' and unsustainable consumerism. They have termed this foundation as 'Style That Lasts'.

These future brand positions are then translated into beautifully realised, new product concepts. Following best practice principles of successful branding, the DNA of the selected brands should be evident in, not just the whole product idea, but every minute decision on form, detail and colour.

In the future, Fred Perry needs to remain competitive in a constantly changing, consumer-driven society. With an emphasis on "Style That Lasts," Fred Perry will need to focus on the emotive attributes of the brand for the well-being of 'Fred People'. Anticipating and exceeding the future expectations of Fred Perry's diverse range of followers will be key to future success.

Wristband Collector

Dan Cherry

The Fred Perry Wristband Collector focuses on the collection of festival wristbands. The primary purpose of these bands is to act as an entry ticket, but they are often worn long after festivals as a display of personal achievement. Their untidiness over time means that they are inevitably removed. The Fred Perry Wristband Collector does not only display used wristbands in a stylish way, but also keeps them secure.

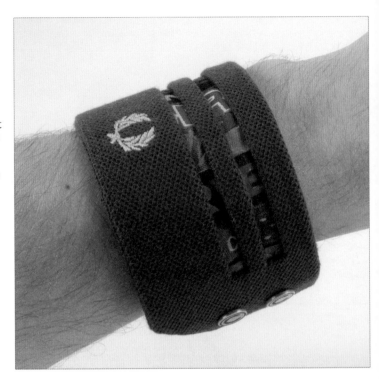

Pocket Projector

David Crittenden

In a world where people are more connected than ever before, and are able to "socialise" anytime and anywhere digitally, it could be argued that, as a digital generation, we are lacking the ability to interact with each other personally. Fred Perry pocket projector allows those who have less self-confidence, to bridge the gap between digital and human interaction.

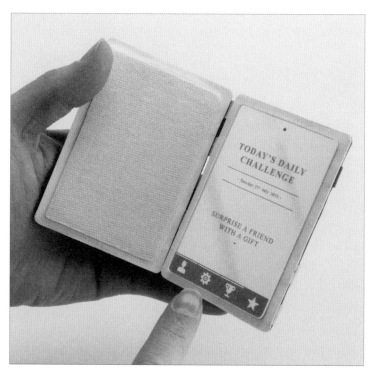

Smart Mirror

Andrew Davies

Many people lack self-confidence and this can stop them from reaching their full potential. Fred Perry Smart Mirror helps to improve short-term self-confidence with the aid of a mirror, and long-term self-confidence through smart mirror technology that displays daily challenges. These challenges encourage interaction with others and increase Fred People's well being, helping them to gain confidence and enjoy life more.

Speaker Headphones

Jack Sandys

Using the idea of well being among Fred People, Fred Perry headphones have been developed to help improve self-confidence and interaction between each other, something so valuable in today's society. The headphones also adjust into speakers for entertaining and bringing people together, making them feel part of the collective experience.

Kinder

Every journey starts somewhere

In 2020, the Kinder brand will place more emphasis on the journey from childhood to adulthood. This new direction will enable the brand to connect to a wider audience, bringing fun and playfulness back to adults. The type of products produced will be based around the idea of bringing back the qualities of youth and the innocence of childhood.

Kinder Cofftea
Dhanish Patel

In 2020, one of Kinder's brand values will be fun and engagement. A child-like quality that is associated with children is the natural fun factor that they exude. As we grow older, adults lose this ability to have fun and become used to performing mundane routines throughout their lives. To solve this, Kinder Cofftea aims to bring a fun, modern twist to making tea and coffee while promoting the idea of bringing people back together to enjoy life.

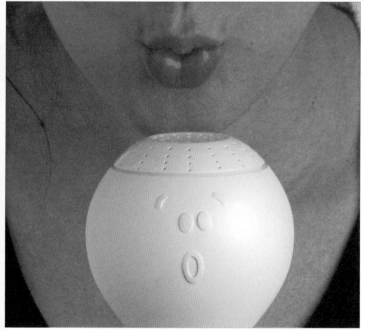

Kinder Mics
Sukhi Assee

In 2020, one of Kinder's brand values will be creativity. Hectic adult lifestyles and daily work commitments leave little time for creativity and fun. Kinder microphones allow users to record their voices and make an array of beats through a simple "vocal tone and shake." Collectively all the microphones combine to produce a selection of sounds, letting creativity flow.

Every journey starts somewhere

Kinder Kinderling

John Aston

In 2020, one of Kinder's brand values will be togetherness. As we grow up, we lose childlike qualities, which helped bring enjoyment and a carefree attitude to life. Kinderling is a unique twist on a fondue combined with a bonfire. This allows people to share food whilst keeping the fun aspect of being around a bonfire. The product helps to rekindle childlike moments of togetherness while the bonfire style ignition adds a ritual aspect to each use.

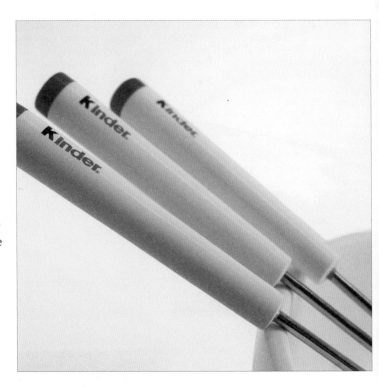

Kinder Massager

Alyssandra Lagopoulou

In 2020, one of Kinder's brand values will be trust. As a child, we rely on our parents to take care of us until we are able to look after ourselves. As an adult, after a hectic day, it is nice to be looked after. The massager allows adults to put their trust in each other to create a relaxing experience.

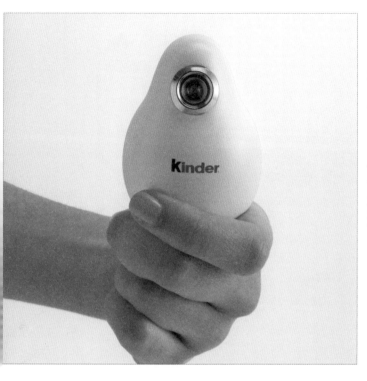

Kinder Kams

Nejal Patel

In 2020, one of Kinder's brand values will be creativity. 'Kinder Kams' are individually different, taking away the precision in capturing moments. Let's start to have fun again; snap away with the 'Kinder Kam' and be intrigued by the creativity you thought that you had left behind. Capture your journey through Pinhole, Lomography and Instagram style images.

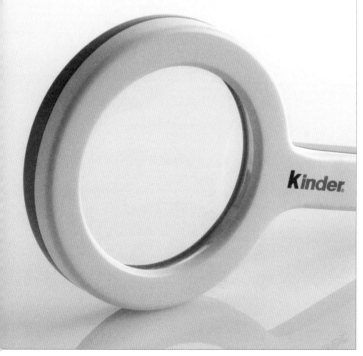

Kinder Curio

George Williamson

In 2020, one of Kinder's brand values will be curiosity. Curiosity plays a large part of childhood but during adulthood it is often denied. Sparking the curiosity of childhood during adulthood, 'Curio' allows for knowledge and inspiration to be found in the conformist yet chaotic world around us.

WWF is highly successful at targeting individuals with an existing concern for environmental issues. However research found that those living in cities, particularly busy professionals, have often become disassociated with nature and its value and therefore donate less. To target this new demographic, a range of products was developed aimed at enhancing professional lifestyle.

WWF:Freetime
Adam Wadey

WWF:Freetime encourages busy professionals to develop a healthy balance between their work and free time. The watchface displays the 9 - 5 working day. At 5pm the hands stop, indicating to the user that work should have ended. The watchframe can then be removed from the watchband and deposited on the watch stand, which displays time for the rest of the day.

WWF:Pinecone
Matt Redwood

The WWF:Pinecone has been designed to be an efficient temperature regulator for the office environment. The design and functionality has been inspired by the way a pinecone regulates its internal temperature when on the tree. Its unusual shape has been designed to promote conversation about its origins.

Getting closer to nature

WWF:Droplet

Barnaby Hunter

WWF:Droplet is a new concept for a water purifier, it is designed to emulate the brand values by providing clean and safe water. The design has been inspired by nature, using the pure shapes of droplets and the biomimicry of organic filtration. This passive integration of nature creates a new style of advertising for WWF with a unique opportunity for a new approach to charitable giving.

WWF:Daylight

Oscar Bowring

WWF:Daylight enhances productivity within the professional environment by replicating natural light. The product incorporates 74 optical fibre strands, illuminated from a 10,000-lux light source. Light falls on the user's face, replenishing their energy, as a preventive measure from winter blues. Environmental values and brand identity are embodied through the use of bamboo, bioplastics and considerations for sustainable design.

WWF:Baseline Beats

Thomas Cody

WWF:Baseline Beats utilises several biomimetic functions to aid busy professionals throughout their daily routines. The honeycomb cell structure works to naturally reduce ambient sounds through noise degradation, allowing the user to concentrate on the tasks at hand. The bull snake's unique epiglottis amplifies sound, a feature present in the ear pad cavities, allowing users to attach their own earphones and relax, listening to their favourite beats.

Positive attitude + positive living

Laziness, poor family values and unhealthiness are some of the most pressing issues regarding young people in the UK. These three concepts aim to tackle these issues in a wagamama way. They have been inspired by the Japanese culture and fused with Western ways to create modern, fun products that would be appropriate and beneficial to young people to improve their way of life.

Wagamama

Positive attitude + positive living

wagamama kizu

Hayley Burrows

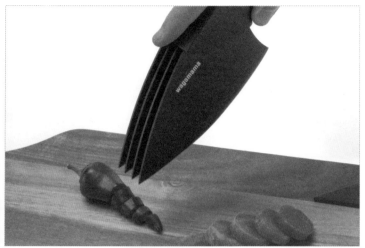

wagamama kizu is an ergonomic knife with four sharp ceramic blades made from zircon oxide. This is the second-hardest material excelled only by diamond. They are 10 times harder than steel blades and will keep their razor sharp edge for many years. The four blades are designed to speed up the preparation of vegetables for fresh vegetable dishes such as stir-frys.

wagamama bibous

Alec James

wagamama bibous are exercise aids that encourage users to move more in their daily lives. Inspired by the look of traditional kokeshi dolls, they are designed to appeal more than the average exercise watch. When connected to an app that tracks exercise on the user's phone, they vibrate to remind their users to do more.

wagamama gobogobo

James Hellard

wagamama gobogobo is a water jug with an internal air pocket that produces a playful glugging sound when poured. The aim of the product is to make family mealtimes more fun. This will encourage families to eat together and build on relationships through social interaction.

Graze

For good health

graze deliver natural and guilt-free snack boxes to busy professionals at work or at home. With a focus on meeting the well being needs of graze customers throughout their fast-paced lifestyles, the concepts encourage and enable them make small changes to their daily routines to improve their fitness, relaxation, personal productivity, and connections with friends and nature.

graze teatime
Edwin Foote

graze teatime is a vacuum flask and tea set that gives co-workers the opportunity to take a break and socialise with each other. With busy work schedules and tight deadlines, it has become more common for people to grab a hot drink on-the-go and spend their working day in relative solitude. By providing a quality tea drinking experience, graze teatime gives a reason to take some time out of the working day.

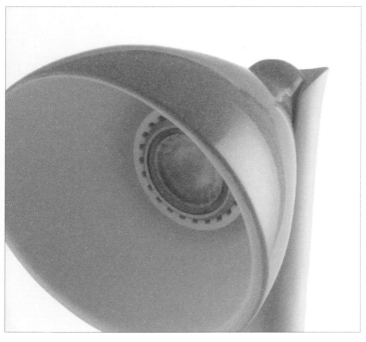

graze illuminate
Chris Naylor

graze illuminate is a desktop lamp that is designed to sit on a desk and make you happy. It provides a way for people to manage their time and home office space for a healthier, happier, and more productive working day. They simply tilt the lampshade and the light is adjusted to their needs - point the shade up and it gives a soft relaxing ambient light perfect for break times; point it down and you get a functional task-light to help you concentrate.

For good health

graze grow

Simon McNamee

graze grow is a modular, hydroponic window farm designed to encourage urban dwellers to reconnect with the source of their food by nurturing their own herbs, fruit and vegetables. The system makes growing your own food easy by delivering seeds and organic nutrients straight to your door. The ceramic pots extend from a central column, in which the nutrients are shared between the plants.

graze active

David Elmer

graze active is a modular water bottle designed to encourage lunchtime exercise for busy professionals. Customers receive liquid concentrate pods of natural fruit juices through graze's subscription-based service. The pods are placed in the flexible silicon base. When the customer is ready to go for a run, they hit the base of the bottle against a hard surface, forcing the concentrate into the water, creating a beautiful effect as the liquids mix.

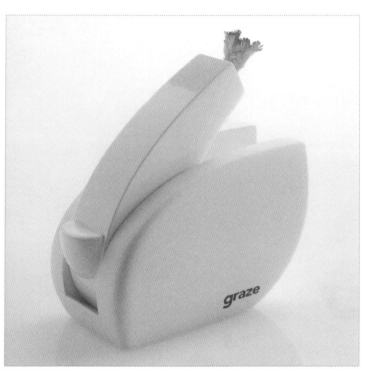

graze relax
Olly Simpson

graze relax is a kinetic, scented oil lamp. The sudden change from a fast-paced working day to sleep means many are unable to relax and rejuvenate before starting afresh the following day. graze relax aims to help ease this transition by building and promoting a positive emotional response through scented oil refills. The lamp lowers to rest over time as you doze off, extinguishing the flame for another day.

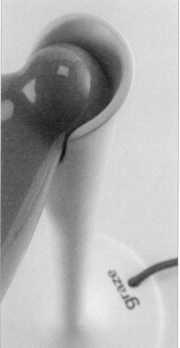

Research into the future of blood donation revealed the ineligibility of four of the six team members to give blood. Give Blood does not rely on the donation of money, so if donors are not eligible to give blood, what else can they give? The future direction for Give Blood is built on the concept that we all have something to give, and the products aim to help people discover this.

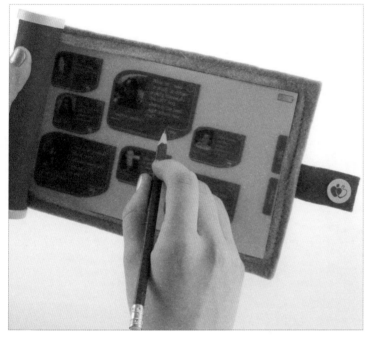

Know-How
James Carswell

Know-How enables people in companies and educational institutions to give knowledge. The aim is to reduce the amount of knowledge lost when people leave institutions, and pass on "know-how" to future generations. With a simple-interface flexible screen, users in the same local network can answer each other's questions, enabling them to build a collection of know-how for future reference.

Supporter
Louis Garner

People do not usually hesitate to offer support to a loved one but when a stranger is in need, people are reluctant to offer support. The product encourages donors to give their support to those in need by connecting them to a suffering person. Donors can subtly monitor the sufferer's stages of grief to get feedback on their progress and to then be able to offer personal support.

Creative Capture
Laura Hodges

Creative Capture is a fun and interactive film-maker for a group of young people working together, either behind the camera or in front of the lens. Working with others in this environment raises self-esteem while honing creativity and encouraging exploration of the local community. It takes all the aspects of professional filmmaking (lights, camera and a microphone) and packages them into balls. They can be easily passed between members during filming in order to capture a new angle or effect, optimising creativity.

People Skills
Sophie O'Kelly

People Skills builds a close connection between a confident Give Blood member and a shy member. Socially anxious members are encouraged to go out to practise and acquire social interaction skills, which are all recorded and discussed with the other member. Having another person listening in can help shy people realise where they are being excessively self-conscious, negative and self-obsessed, hence improving their interpersonal skills.

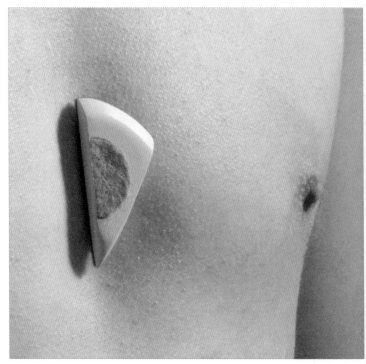

Time Totem
Sarah Hutley

Time Totem is given to blood donors on their last possible donation. With the product, donors can continue to be a valuable part of Give Blood even though they are no longer eligible to give blood themselves. Time Totem presents people with giving opportunities in their local community enabling them to give their time helping others.

Experience It
James Ward

Experience is a personal and irreplaceable part of life that we constantly develop and expand upon, no description or array of visual media can replicate the actual experience. We can only imagine based on our own real life experiences. Experience It changes this by replicating not only the exact visual imagery, but also to a degree, real-time senses, emotions and bodily reactions as experienced by the capturer. The resultant effect is the formation of a true connection between two people sharing a personal experience.

Let's do it together

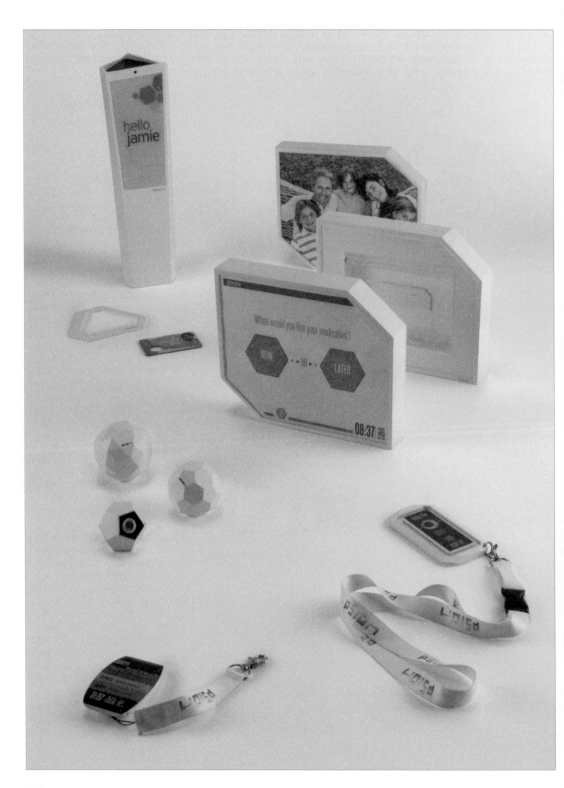

The key to Psion's success in the world of 2021 lies in providing the brand with an energised, passionate and engaging character. This will provide an enhanced level of personality and community, which consumers are able to relate to on a more intimate level. The personality inspired a simple, playful and coherent language across the products.

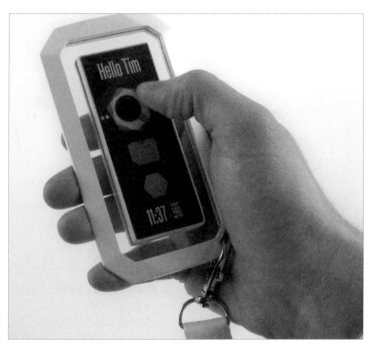

Psion Events Pass
Tim Logg

Events such as trade shows allow us to exchange information and stay connected. From exchanging contact details to downloading exhibitor information and way finding capabilities, Psion Events Pass is a complete personal aid for anyone visiting a variety of events. Information can be downloaded directly onto the cloud or the visitor's personal electronic device allowing the information to be accessed anytime.

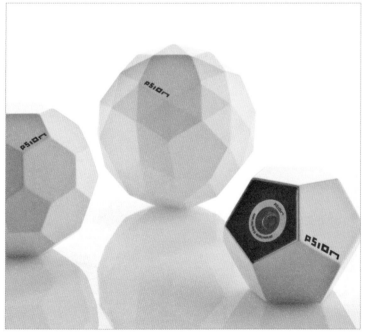

Psion Sidekick
Luke Gray

The Psion Sidekick is a tool to help improve and simplify a student's experience in education. The device allows for simple access to interfaces and devices within the school and acts as a physical link to a school's community and online network. Other features help the user organise their student life and school work in and outside of school.

Let's do it together

Psion Stream

Joe Carling

Psion Stream can be picked up on entering a shopping centre. It enables users search for items using real-time stock data and route finding. Items are added to your "shopping basket" by holding the device next to the object and can be organised later before paying as the device is returned. In this way shopping bags don't need to be carried all day and on paying they can be delivered home or straight to your car. This allows multiple shops to store their stock centrally, allowing for more efficient storage and space optimisation.

Psion Touchpoint

Jamie Topp

In 2012, Psion Touchpoints will be situated around the community, allowing people to plan every step of their journey on any type of transport through a simple, friendly, easy-to-use interface. Psion Touchpoints recognise every person registered with the service allowing for a more personal experience. Psion Travelcards allow people to plan ahead for the whole of their journey as they go.

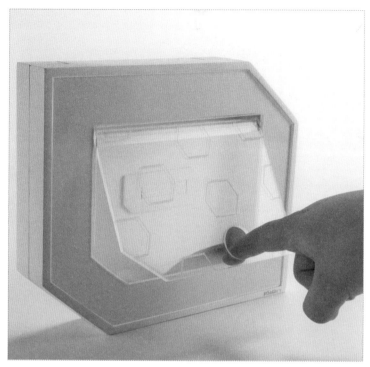

Psion Healthcare
Deborah Wrighton

Psion Healthcare allows patients to control and manage their personal medication in their own home. A wearable reminder alerts the patient to take their medication and directs them to the three wall mounted units. The medication, which is dispensed on a bite-size rice paper slip, can be taken instantly or stored to take while out and about.

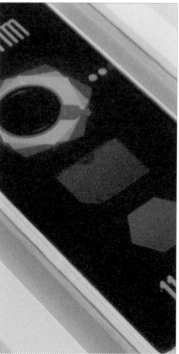

Cashless money management

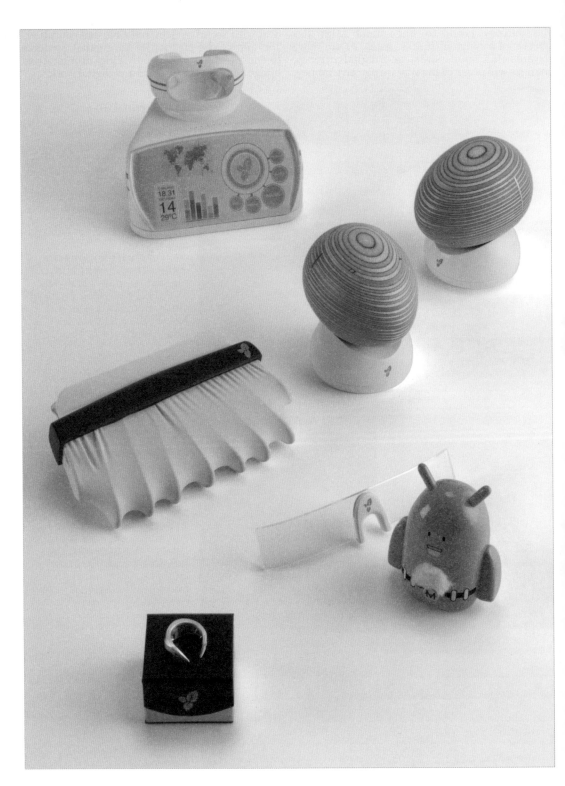

The journey taken to understand Mint has uncovered a range of future issues. As we move towards a fully immersive stage of digital currency, changes will occur in the way we perceive money, deal with security, debt and financial education. The aim is the creation of conceptual products to address these issues, providing an interactive bridge between the physical and digital world.

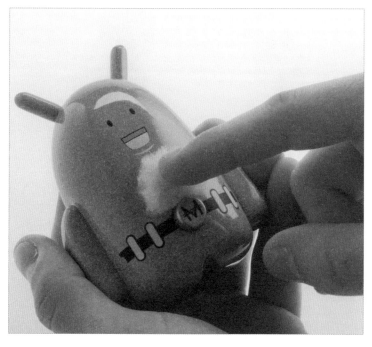

Little Minter
Cameron Henderson

Personal debt is reaching an all time high and financial education is an absolute necessity. Introducing "Little Minter," an augmented reality adventure. Throughout the journey, Little Minter offers guidance on how to utilise coins earned, providing the child with a relevant financial foundation. Additionally, by directly linking their financial situation to Little Minter's health, a stronger sense of responsibility is developed.

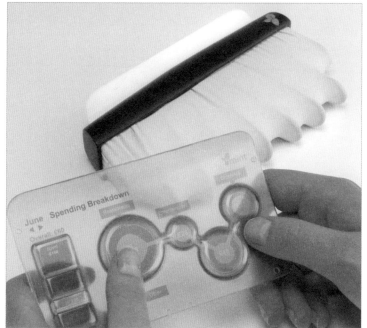

Clutch
Ben King

Clutch is a dynamic electronic payment device for 2026, where digital transactions are the norm. Mint supports users by physically representing their spending through a tactile interface. The shape of Clutch transforms in response to the level of funds available on the device, with the pull out card offering a detailed control of your spending.

Cashless money management

Mint Pinch
Dimitrios Stamatis

As we get closer to a time when money will no longer be represented as physical objects but as electronic information, a perceptual shift will change the way we acknowledge our personal finances. Mint Pinch is a ring, which when worn, can alert the wearer if their account is being hacked. The concept has embedded connectivity and a response system based on pain feedback, which has a higher travel speed than any other tactile signal.

Mint Guardian
David Johnston

In the UK, approximately 250,000 students take a gap year before university. Mint Guardian aims to help keep their finances on track. With a watchful eye, parents receive timely alerts for unusual account activity through their home device. The student stays sentimentally connected wherever they are in the world through their wristband while being notified of any important issues.

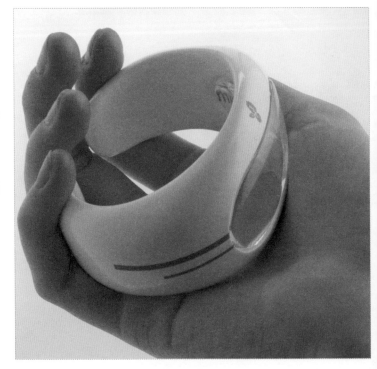

Mint

Mint Exchange
Sam Edwards

To bring back the tactility associated with cash, this biometric payment device is used by consumers and retailers. Its design details reflect qualities found with physical money, with layering and metal edges owing their origin to cash. These materials wear with age, and bring familiarity to those who are apprehensive of a world without physical money.

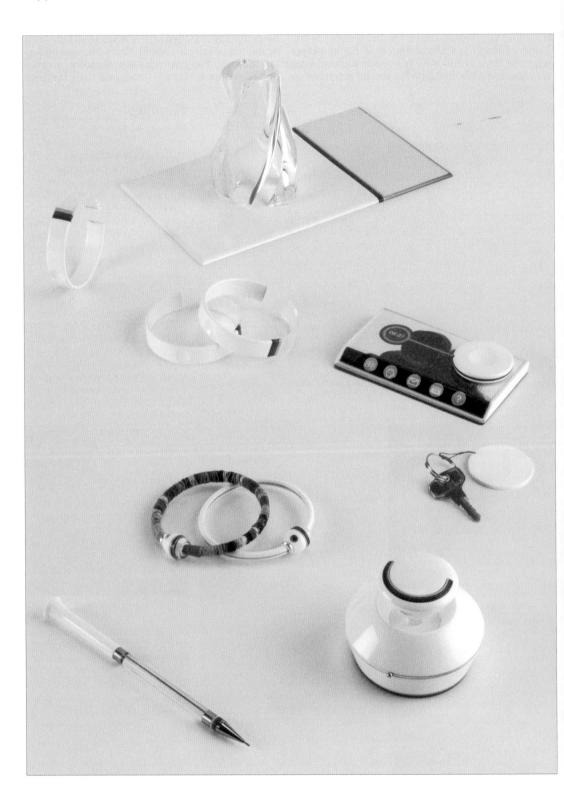

Help for Heroes
Support for our wounded and their families

It's about the 'blokes', our men and women of the Armed Forces. It's about Derek, a rugby player who lost both of his legs. It's about Richard who was handed a phone as he lay on the stretcher so he could say goodbye to his wife. It's about Ben, Andy and Mark, it's about them all, our blokes; our heroes. Help for Heroes, in 2020, aims to provide support for our Heroes and their families.

Families
Patrick Bion

Working to help families as they reintegrate together during post-deployment, this product aims to subtly encourage greater communication between family members. Wristbands, worn by each person, measure the individual's fatigue and stress. These measurements dictate the movement and growth of a visual representation of the family, showing individual well being and the integration together. Through this gentle visual representation, the family can continually adjust what responsibility is placed upon one another, and help them to be more aware of each other's well being.

Military Women
Sophie Randles

The pen aids female soldiers to combat the negative feelings and emotions experienced when working in a male dominated environment through interactive tattoos. When the soldier replicates certain patterns or words with the pen, a tattoo linked with that pattern appears on her skin. This acts as an empowerment tool to give her confidence and reassurance. The tattoo fades as her belief in herself improves.

Partners

Emily Riggs

The product is designed for military partners, to assist them both when their partners are at home or away. When away it will enable them to feel that they are being left in safe hands. Military wives interviewed felt they relied on their husbands passing on information and wanted something that could be given directly to them. This is already in place but some are not aware of it. This product improves the connection between users and information.

Relocated Children

Victoria Gibson-Robinson

Part of growing up in a military family is frequent relocation. Children who have parents in the Forces move according to their parents' assignments. Memory bracelet enables children of those in the Forces who are frequently relocated, to carry their memories with them. The bracelet records the colours of the child's surroundings throughout the day. Twisting the bead on the bracelet prints and weaves a thread around the bracelet, reflecting the colours of the child's day. Over time, the bracelet is filled with colours, which represent the child's memories, allowing them to carry their memories with them wherever they are.

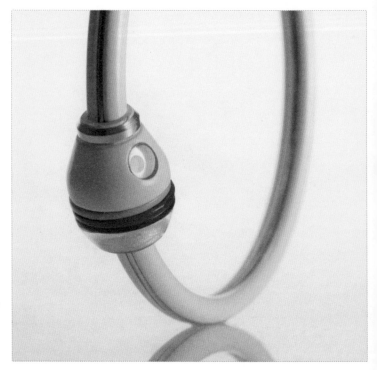

Children
Emily Menzies

This product enhances the interaction between deployed military parents and their children. It provides a daily base of contact through fun and exciting methods of communication. Children can unlock new games, send and receive messages, and share photographs on a daily basis. The product visually represents the length of time until the parent returns home, helping to maintain morale.

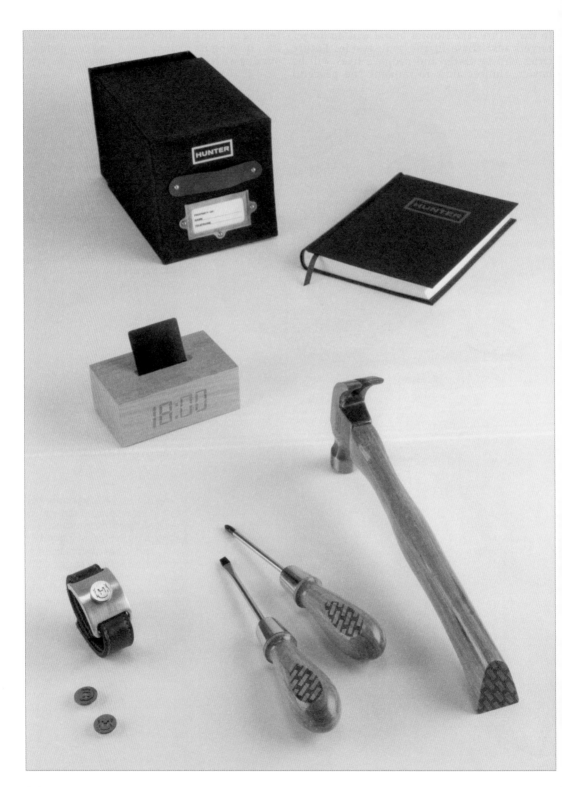

Hunter

In the future, the Hunter brand will no longer targets just those in the countryside. Future cities will be faster and people's lives will be busier. Hunter aims to implant the peaceful feeling of living in the the countryside into city life, helping us to slow down and complement our day with countryside emotions

Victoria Wristband

Phil Verheul

The future will see an increase in the number of city commuters travelling to and from work every day. Victoria Wristband helps tackle the busy commute with its interchangeable pods that exude a smell. When breathed in at close range, they will help to transport the user's mind to a calmer place.

Charing Hardback

Cianan O'Dowd

In the digital future people will be dependent on the screen for all of their information. People like the feel of a good book, and it has always been the traditional way to get information and organise photos. The Charing Hardback aims to detach us from our digital screens by sharing our photos the way they were meant to be shared.

Oxford Box

Jesse Williams

Living away from home and family can be difficult and sometimes digital connection is just not enough. The Oxford Box is a ration box from family members. It contains items that they have chosen themselves. These items connect you to your family when they cannot be with you physically. The Oxford Box provides comfort and flavours of home by letting you know your family is thinking about you.

Westminster Clock

Ben Crichton

Westminster is a digital clock with a case hand-crafted from multiple sections of English oak, giving the illusion of a seamless finish. In a world where people are increasingly taking work home with them, this charismatic product promotes a better work-life balance through haptic, visual and auditory stimuli.

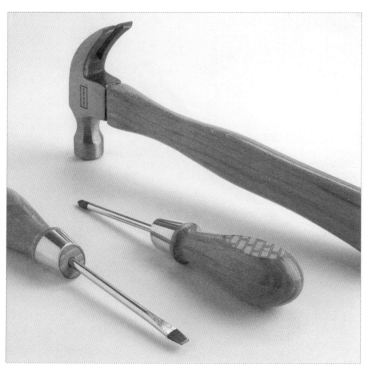

The Sussex Tool Set
Roland Skinner

Sussex Tool Set brings together the perfect tools for the job. Using high quality materials and classic styling, each crafted tool instils the warm feelings of home. Modern tools often use cheap plastics and rubber in their construction, while aiming to look good they often fall short. Hunter tools use natural materials which not only look good but last. Using the Sussex Tool Set instils a feeling of craftsmanship, pride and well being.

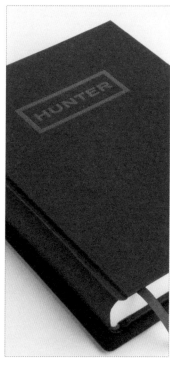

Creating beautiful moments since 1863

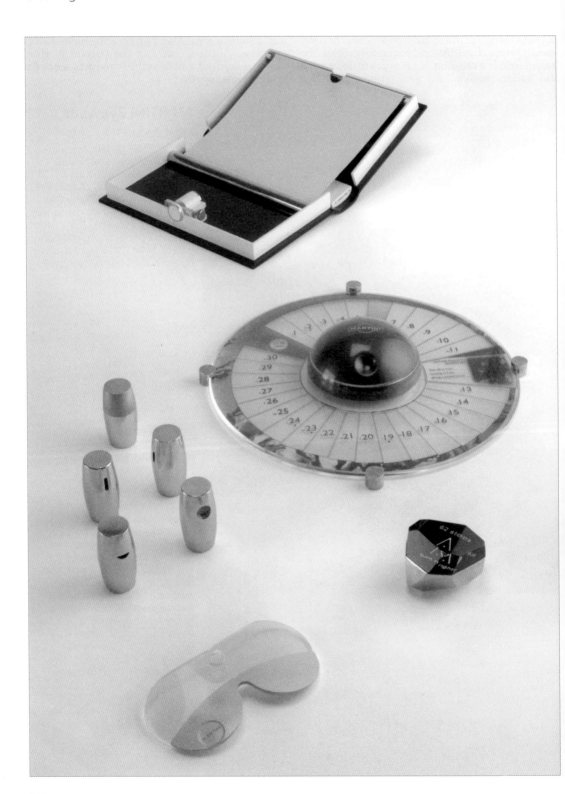

Martini
Creating beautiful moments since 1863

MARTINI is one of the world's most iconic brands, known for its charisma and beautiful drinks. The Italian icon is a creation of 140 years of dedication and Italian passion. For generations, MARTINI has brought people together, creating memorable moments. With a focus on the year 2026, this range of spirited and beautiful products seeks to bring people together.

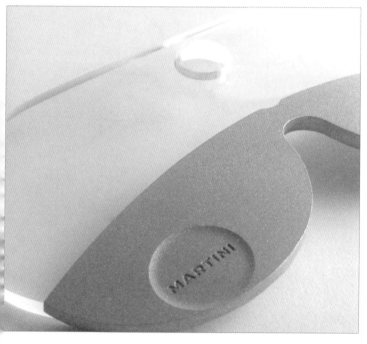

MARTINI Eyewear
Luke Bacon

MARTINI Eyewear enables two people in an intimate relationship to stay connected. Information shared between them is displayed on the lenses within the eyewear. For a more intimate connection, only one pair of MARTINI Eyewear can be connected at any one time.

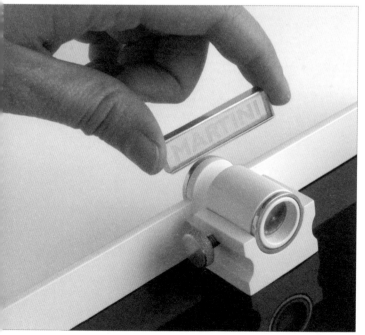

MARTINI Pictures
Sophie Richards

MARTINI Pictures provides a hub to relive and share memories through photos and videos. The product allows you to add comments and details and even arrange another moment to add to your collection. Look through your memories, randomly or by event, triggering those all-important reflections. Sit down, pull out the canvas, insert your memories, relive and share your moments.

Creating beautiful moments since 1863

MARTINI Diary
Matt Bastow

MARTINI Diary is devoted to rekindling ambitions from the past and ensuring that future aspirations are achieved. The interactive diary allows you to plan for retirement life through automatically and simply inputting your ambitions and dreams, providing a simple and intuitive way to learn more about your plans, and ultimately making your dreams a reality.

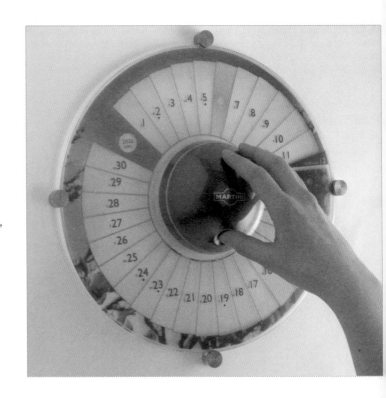

MARTINI Senso
Laura Ginn

MARTINI moments are some of the most precious memories in life. The most prominent memories are often evoked through the five senses, transporting you back to those moments. The precious MARTINI Senso collection allows you to capture each of your senses, to later playback and relive the whole experience, as if you were having those MARTINI moments once more.

MARTINI Conscience
Mark Taylor

Life can be full of regrets, something we are all familiar with. Out of the five most common regrets, four could easily be prevented. This crafted product simply provides you with confidence, pushing you in the right direction towards achieving goals and discovering new relationships. The design is as unique as you are - no two forms are the same.

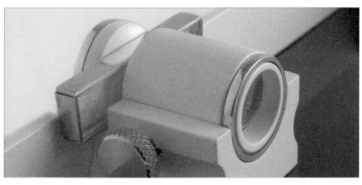

Quaker Oats

Future family togetherness

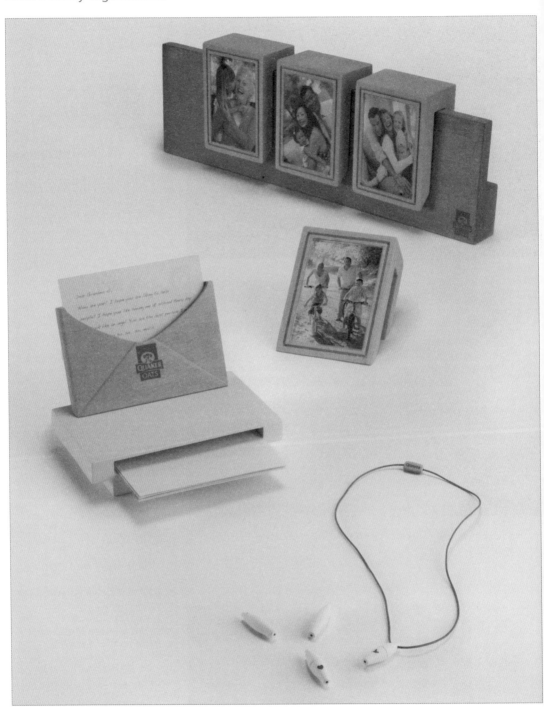

Quaker Oats is deeply rooted in integrity, honesty and purity and is the brand of choice for hot cereal and nourishment for families, enabling them to share memorable breakfast meal times. With the world changing, families living apart, Quaker Oats is faced with the threat that globalisation poses to its ability to keep families together. By 2025, Quaker Oats needs to find new ways to keep families together.

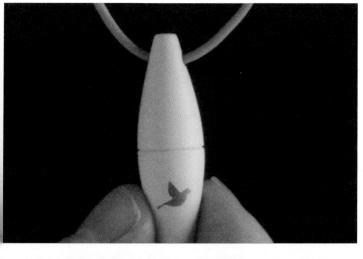

Family Seeds
Vincent Ashikordi

Family Seeds are symbols of family unity and togetherness. They contain family values and memories passed down from previous generations. Each seed serves as a rich source of strength and moral support and can be plucked from the family tree to be worn as a pendant when family members are apart, enabling them to cope in times of loneliness.

Family Recipes
Jaymini Desai

Family Recipes is a product that helps people to recreate their favourite family recipes when apart from their loved ones. This future concept for Quaker Oats functions as a recipe storage and display system. Hidden behind the family photos are recipes that can be accessed using the touchscreen display.

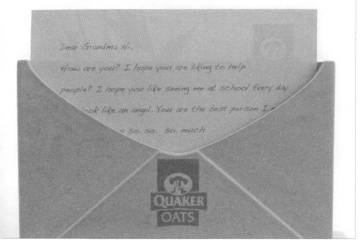

Telex
Archit Rakyan

Custom online socialising provides a good opportunity to stay in touch, but receiving and reading a warm message in the morning is more personal and an equally important ritual as having a family breakfast. Telex provides a strong and closer connection with loved ones, enabling you to nourish important bonds just like your own health and well being.

Sunseeker
Traditions are timeless

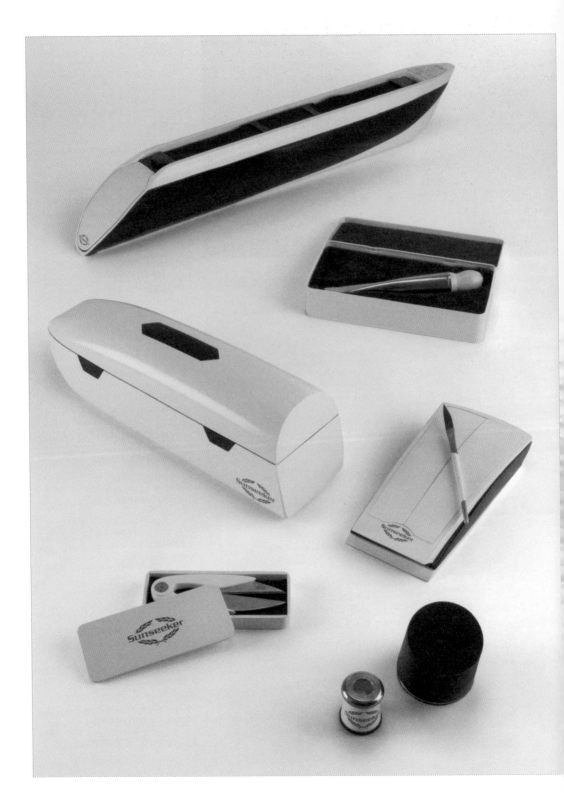

Moving Sunseeker into the product market, these products, reduced in size and cost, retain the same high levels of quality as their continued yacht production. The future direction facilitates the passing down of knowledge from one generation to another, keeping craft skills alive, to be passed on to future generations.

Precision Pruner
Peter Chueng

Precision Pruner is a luxurious garden tool that reflects the prestigious values of the brand. It combines ergonomic comfort with high quality craftmanship offering you a pleasurable and personal experience. With Precision Pruner the craft of gardening can be passed down through generations.

Drawing Tube
Exequiel Di Salvo

Drawing Tube has Sunseeker's values, craftsmanship and modernity in a neat, personalised package. The tube allows you to hold two scrolls of personal drawings, protected by a yacht-inspired slider cover. A retractable strap allows you to carry the product on the shoulder and when not in use it is slotted flushed on the tube for a streamlined appearance. This product helps to realise the dreams of those who appreciate perfection and provides an emotional bond through a shared artistic activity between generations.

Sunseeker

Traditions are timeless

Wine Tote

Jacob Lee

Wine Tote is a high quality carrying case for a bottle of wine. Wine tasting is a skill that is as old as it's production and still continues to be passed down today. Designed to last, Wine Tote will be passed down from generation to generation.

Brush Strokes

Sunil Patel

Painting is a skill that is learnt and developed through experimentation and practise. A selection of 5 brushes, all designed to create different brush strokes, are encased within an easy to access unit. The unit also stores the essentials needed to maintain the brushes, ensuring the same quality and skill is consistently passed down from generation to generation.

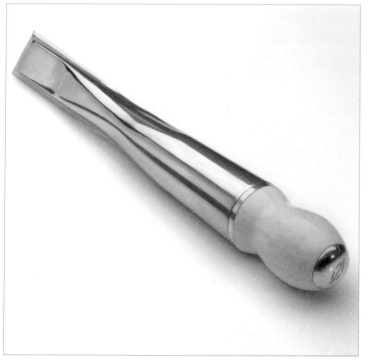

Bevel Edge Chisel
Richard Sheppard

Bevel Edge Chisel is a stylish and highly functional tool, ideal for the professional or keen amateur user. Crafted from timeless materials, the tool can be passed on to future generations, keeping the traditions and craft of wood working alive.

Optical Loupe
Oscar Correa

Optical Loupe is a jewellery magnifying glass which can magnify from 3x – 6x optical zoom. It is crafted with timeless materials and exquisitely finished. Inspecting materials and the details of products is a great way of educating a child and showing them value and craftsmanship. This product is about passing down a skill with Sunseeker's prestigious, sleek and technologically driven styling.

Eddie Stobart

Keeping your family moving

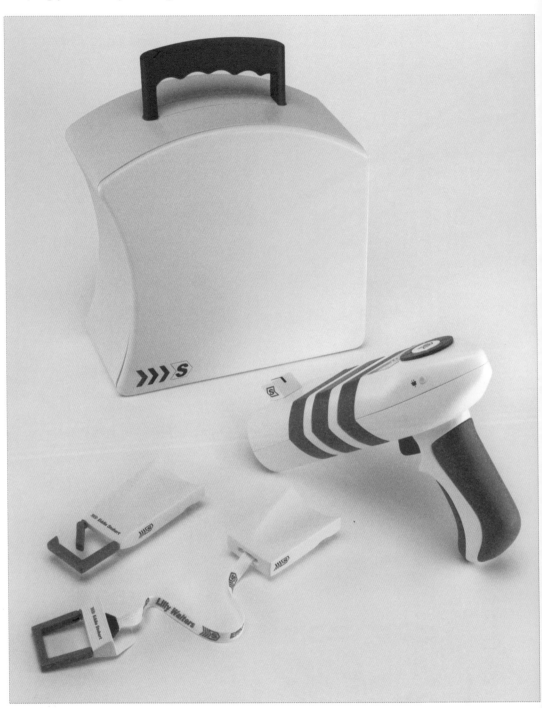

Eddie Stobart is an iconic British brand, rich in family values. An increasing pace of daily life presents the possibility of families growing apart. Eddie Stobart products will keep families of the future together. From connecting children to their working parents, to a more efficient morning rush, Eddie Stobart products will be there every step of the way.

Eddie Stobart

Keeping your family moving

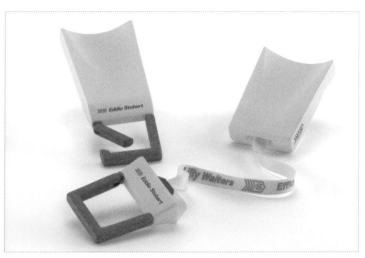

Efficient Eddie

Jon Taylor

The morning rush can be a nightmare for everyone, especially parents, as school children seem to leave everything until the last minute. Efficient Eddie is an interactive daily organiser designed for school children to address this issue. With the device, school children can prepare in time for school, get homework and PE kit reminders, and view their daily timetables.

Cool Eddie

Jamie Smith

As family life becomes more hectic, preparing for the early morning rush becomes more challenging. Cool Eddie alleviates the stress imposed by preparing breakfast before work and school. With the lunch box, breakfast prepared the evening before for loved ones can be safely stored in temperature-controlled compartments, saving families time and relieving stress.

Create Capture Share

Salvatore Bella

Create Capture Share is designed to help parents rebuild their relationships with their children, which have weakened over time due to busy lifestyles. With the product, a child can create work they are proud of, capture their work through a 3D lens and share it with their parents for feedback. This encourages frequent interaction and improves the relationship between parents and their children.

Tefal

Healthy eating driven by technology

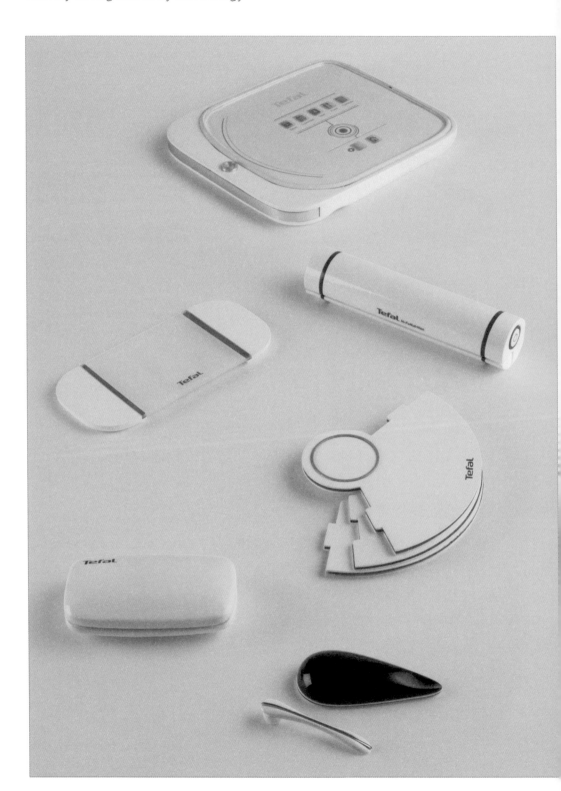

Tefal is dedicated to understanding consumers' needs and delivering innovative solutions. Tefal's history is marked with world-class innovations that communicate its brand identity.

Tefal's brand values in the future should be up-to-date, often trendy, blend into your life and home, stimulate desire and easy to use.

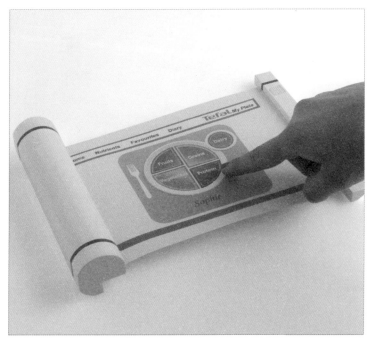

Hi-FuNutrution

Shazwani Shamsuddin

Hi-FuNutrition is an interactive device with a healthy eating guide feature. It allows you to find out about their recommended daily eating plan, menus and recipes using wireless internet connectivity. It enables you to find and read recipes, and buy the ingredients online. It also features entertainment programs, making being in the kitchen more enjoyable and fun.

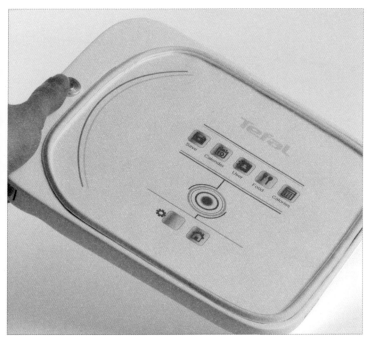

Touchplate

Jason Singh

Touchplate provides interactivity to track your eating lifestyle. With its touch interface screen you will be able to monitor your calorie intake. It incorporates finger print scanning which recognises you and then presents nutritional information catered to you. Keep track of goals, eating habits and have suggested eating regimes to maximise your lifestyle.

Balanced Diet

Stefani Karaoli

Balanced Diet is a portable system that is able to scan and display information on sugar, calorie content and the value and benefits of a food product on the screen. The system contains a list of meals necessary for daily nutrition intake, enabling you to make healthy eating choices.

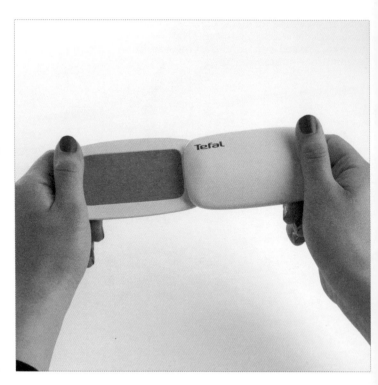

IntelliCook

Yunqian Du

IntelliCook is a portable hob for future kitchens. Future Kitchens will have fit-in technologies like wireless electricity and touchpads in most plane surfaces. Tefal IntelliCook allows you to control heat output from your hob by hand gestures. Drawing a circle on a plane surface regulates the temperature of the hob. The system has features like intelligent pan detection and preventative boil-power.

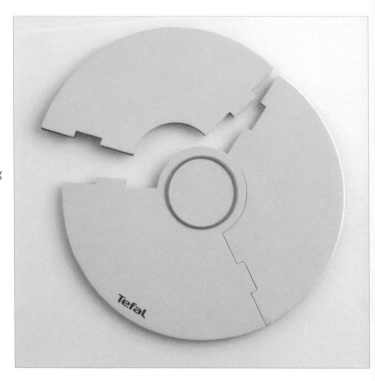

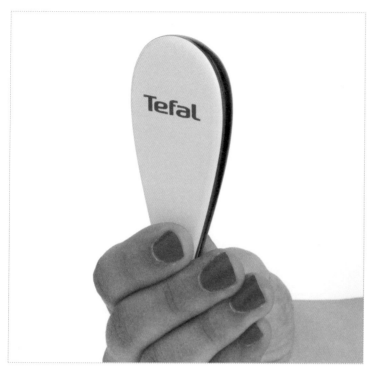

Shopping Assistant

Maryam Abdul Elahi

Shopping Assistant uses RFID tagging system to locate healthy food that has expired in your fridge. It helps you make payment without stopping at the till, through automatic deduction straight from your bank account. It updates, displays and interacts through gesture technology, holograms and additional intelligent earpiece which recognises the aisles making shopping easier for you.

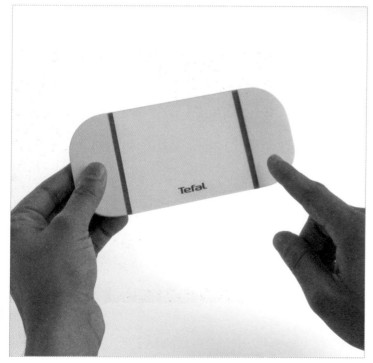

Growpad

Tajinder Cheema

Growpad combines the knowledge of a seasoned food grower and motion sensing touch screen technology. With growing population and food shortage fears, Growpad aims will encourage families to grow their own food from seed to harvest. Households can test their garden soil to activate the Growpad gardening mentor feature, helping them to begin their eco-friendly and money-saving journey to self-sufficiency.

Design Thinkers of Tomorrow

How would they be nurtured?

There is a growing interest in the idea that design thinking provides businesses with the creativity and innovation that they need to compete in the current economic environment. In a report by Vanessa Wong of Businessweek, she looks at design thinking as a creative technique for businesses, for which there is no consensus on how to teach its methods.

> *"Design thinking is the use of design processes and methods to foster innovation and grow businesses."*
>
> *Rhymer Rigby*

Given the global financial crisis, many students are now graduating with the prospect of chasing fewer jobs in the global labour market. Once, there was a constant demand for fresh new graduates who excelled in their given academic disciplines. The whole story has changed. With companies looking to out-compete their business rivals, those graduates who can provide the competitive edge, are versatile, are emerging from interdisciplinary programs that integrate design, technology, and business would be the ones given more attention argues Wong. It is believed that these new graduates are trained as "design thinkers" and it is hoped that this new breed of designers would help provide creative solutions to some of the present day global economic challenges says Wong. But the limiting factor to design thinking is the lack of consensus on how it is, or will be taught, and where future design thinkers will emanate from. Perhaps the reason behind the lack of consensus stems from a lack of understanding of its meaning. Some experts ponder if there is a dichotomy between design thinking and design. Others think design thinking is an extension of the design discipline in which there is an appreciation for the methods used in other disciplines. If in some countries, design is still not regarded as an academic discipline then the hope for consensus is still far-fetched.

Design thinking is a new discipline and just like any discipline in its infancy, understanding its meaning and applications can be challenging. It is, says Rhymer Rigby, author of "T: one day all designers will be this shape" for the Design Council, "the use of design processes and methods to foster innovation and grow businesses." It is a term developed at Stanford University, which builds on theories around creative culture and thinking styles and deploys design methods within strategic business management.

Advocates of the design thinking discipline believe that the theory: Working across functions will offer new perspectives in creative innovation in the corporate world. "The aim is to combine creative confidence and analytical ability", says David Kelley, founder of Stanford's d.school and design consultancy IDEO. He goes on to say that "the best students are competent in both."

The problems facing Western businesses are too interwoven to be tackled in narrow-mindedness. They require a multidisciplinary approach to problem solving, a marriage between technicality and "creative freedom". It is a method that has been proven to work as revealed by Apple's former CEO, the late Steve Jobs, during the Apple WWDC 2010 Keynote Address, that, "the reason that Apple is able to create products like the iPad is because we've always tried to be at the intersection of technology and liberal arts, to be able to get the best of both."

There are many companies today that are great at solving highly complex problems but poor at solving more ambiguous ones. As identified by the Economist in New York last year during their Human Potential summit, to "examine why one of the greatest challenges for the next decade will be attracting and developing hybrid thinkers who can cut across traditional business silos," in such companies, when the goal is unclear, the gears grind to a halt. For such companies to survive in an uncertain world, they need to attract a different kind of talent: multidisciplinary people who combine together previously unrelated fields.

In Bruce Heiman and William Burnet's report: "The Role of Design Thinking in Firms and Management Education," they suggest that design thinking requires multidisciplinary teams; fluency; experienced-based, user-centred research; prototyping; iteration; critique; and form-giving to be successful. In addition, nurturing design thinkers of tomorrow requires embedding a new type of design culture at both the institutional level and at the corporate level.

The constituents of the design culture should embody a multifaceted approach that is centred on visual communication and focused on problem processing activities required of an academic discipline. The thinking here is that if tomorrow's design thinkers are to be able to design solutions to broader societal problems, they have to be able to work across functions. In multidisciplinary teams, they will be at the centre of articulating knowledge, skills and creativity through but not exclusive to effective visual communication and information processing activities such as brainstorming; project mind mapping; storyboarding; and concept generation and development. To be able to achieve this, a new design thinking curriculum is needed, one, which emphasises critical and problem solving capability, autonomous learning and self-initiated learning requirements with the delivery of the curriculum being through a learning environment that is a reflection of the culture of the design institution argues Clarke A. et al in Post Modernity: Design Education Culture, and that of the collaborating business or corporation. The curriculum should encourage social interaction among students from different disciplines within a faculty.

Business, technology and liberal arts should be integral to the new design thinking curriculum to provide students with the commercial awareness, creativity and technical know-how they need to design commercially viable products and creative services.

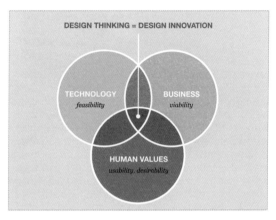

In Wong's report, she asks if design schools should create more business-focused creatives, or if business schools should foster creative thinking in their MBAs. Institutions currently doing so are listed in the 2012 Businessweek D-school List. "These schools are responding to current trends and are taking advantage of their unique strengths to formulate new curricula."

But there is no other place in the West where design thinking is making more progress as the US, perhaps due to the requisite nature of their MBAs and the growing need for their business schools to adopt design methods. "Where to send managers to learn how to be creative is becoming an important issue for top executives. In a business world where creativity and innovation are at a premium, skills in administering organisations have less value," says Jennifer Merritt and Louis Lavelle of Businessweek, in response to the idea that "Tomorrow's B-schools might be D-schools."

To conclude, Wong adds: "Designers who exhibit business acumen can be involved at a more strategic level within a corporation. Executives who learn to apply design methods such as prototyping or brainstorming have a better shot at building a corporate culture that nurtures innovation—and the business' bottom line."

Vincent Ashikordi

Product Design Engineering

NFC Scan 'n' Pack

Future personalised shopping experience

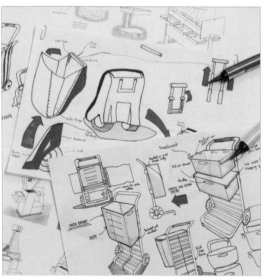

Research shows that based on current trends, by 2016, 85% of point of sale terminals will support Near Field Communication (NFC) payments and by 2017 the retail industry will be making significant use of NFC for customer payments. This project involves a reconsideration of the current self-service checkout solutions, incorporating NFC payment methods, from the perspective of a shopper. The future of retail will be based around personalised experiences, multichannel shopping and intuitive interfaces that are not currently fully developed. This project was carried out with NCR, the market leader of self-service checkouts and ATMs.

Dhanish Patel

Industrial Design and Technology

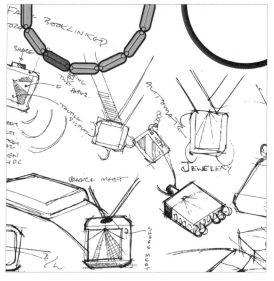

The VIBE platform presents the opportunity for a new music chart based on the beats that get festival and nightclub goers moving. The motion and movement of attendees are tracked against the songs, events and artists, and used to map their VIBE. VIBE takes music events online, offering people real time feedback in the search for the heart of the party. Following an event, data is shared through social platforms, offering a valuable tool for the entertainment industry. VIBE evolved from human centred research into the lifestyle and party habits of 'tomorrow's generation'. The VIBE Platform was the result of a collaborative project with Sony's Design Centre Europe, supported by Sony's Alsace Technology Centre in the development of working prototypes.

Ben King
Industrial Design and Technology

243

ChocoPrint

A fun and intuitive 3D chocolate printer for children

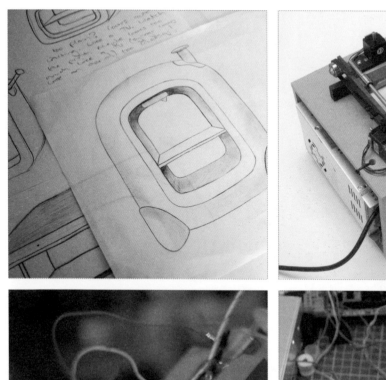

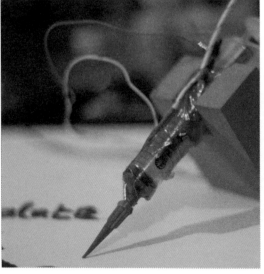

'Chocolate & Confectionery' and 'Toys & Games' are the two biggest markets for children in the 7-12 year age range. ChocoPrint aims to target both markets by taking chocolate feedstock and repurposing it into 3D chocolate forms. Using the Fused Deposition Modelling process of layering material sequentially, chocolate is extruded through a temperature-controlled pump, creating a 3D model layer-by-layer. Advanced methods of heating, pumping and positioning are designed to be cheap to implement whilst retaining much of the accuracy and reliability that industrial Computer Numerically Controlled machines boast. The resulting mechatronic platform that is both inexpensive to produce but fun and intuitive to use.

Tom Cody

Product Design Engineering

ChocoPen

A fun and intuitive 3D chocolate pen for children

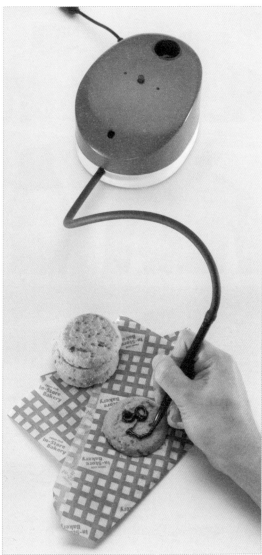

Aimed towards children, this chocolate pen will allow for creativity with an already favoured food material, chocolate. Through accurate controlled heating (PID), chocolate is melted and pushed through a uniquely designed multi-channel peristaltic pump and delivered to the tip of the pen. The flow of chocolate is controlled through inputs on the specially designed pen. Designs can then be doodled onto aluminium foil to be set, shown off and eaten! The hardware and electronics will be designed and built into a working prototype, where written code will instruct the hardware how to operate.

Tom Cody

Product Design Engineering

Pilot Survival

Survival and recovery suit for light aircraft pilot

This project is a feasibility study of the design of a suit that is to be used by pilots in a forced landing. Its main purpose is to enable the pilot to survive and to recover in such a situation. Statistically, these incidents end with injuries, and some with death. Research was carried out on how to tackle this problem, which included new and existing technologies such as power harvesting and smart materials to assess their suitability. This led to the design of a fully integrated suit that counters threats faced by a pilot in many environments. This survival suit, if realised, has the potential to save lives in what could otherwise be a deadly situation.

Andrew Matthews

Aerospace Engineering

The project objective was to retrofit a model aircraft with Morphing wings and no conventional roll control surfaces. A variable (retractable) wingspan was chosen, with warping wingtips for roll control. This idea was chosen as it had the potential to offer multiple performance increases compared to the baseline aircraft. It should be able to increase the maximum speed, manoeuvrability and endurance of the aircraft. When the wings of the aircraft are retracted, the increase in the aircraft's performance compared to the baseline should be noticeable.

D. Pollard, S. Phillipson, J. Starley, P. Smaridge, S. Williams
Aerospace Engineering

catch me!

Intelligent feline nemesis

catch me! is a dynamic electronic toy for cats. Styled to imitate the classic feline foe - the ball of yarn, the toy uses RFID technology to sense where the cat is. As the cat approaches, the toy can tell that the cat is coming to get it, and uses its enclosed inertial sphere mechanism to roll away, providing a challenging and intelligent foe to the perplexed cat. Now that's fiendish!

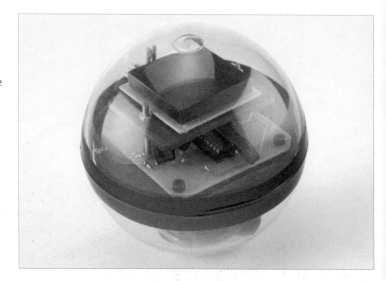

David Elmer

Product Design Engineering

RFID Cat flap

A smart cat flap displaying the position of your cat

An increasing number of cats are being implanted with RFID chips in case they get lost. This cat flap design aims to utilise this chip to allow or deny entry through a cat flap, ensuring that stray cats do not enter your house. These chips will also be used to display the wherabouts of your cat, either inside or outside the house, and the time their status last changed.

Emily Menzies

Product Design

Our Network of Contacts

To produce validated and considered designs, we were encouraged to seek a diverse range of expert professionals within respected fields. This infographic depicts the centre of our extensive network of contacts, their significance and their interrelationships.

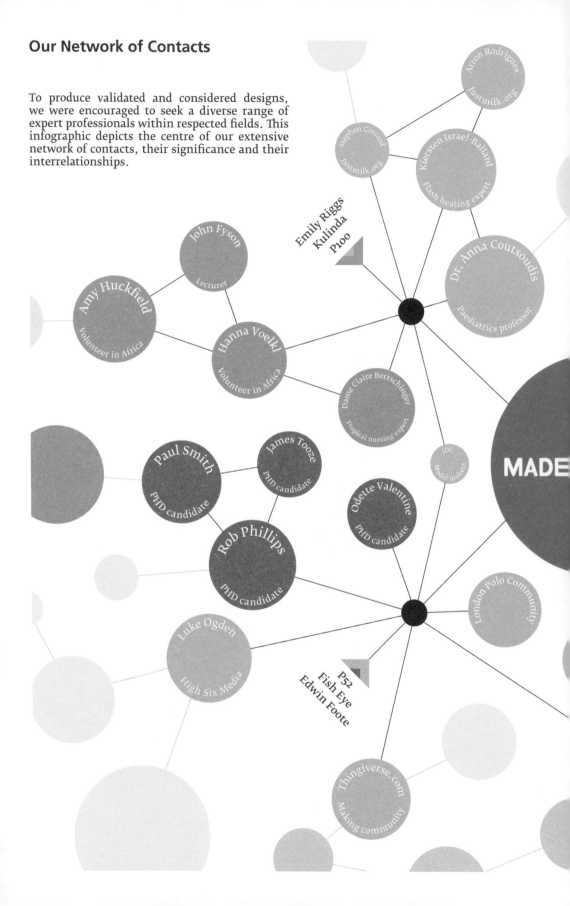

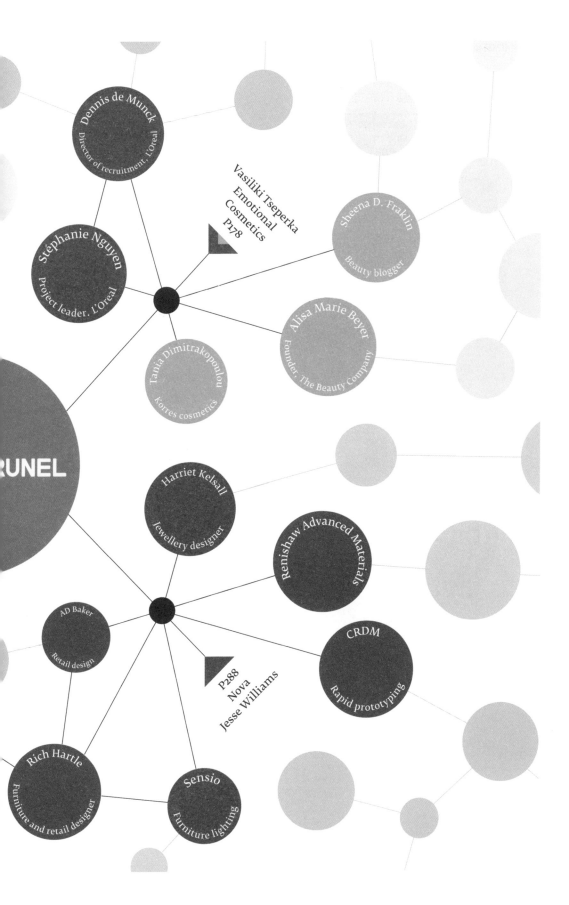

RUNEL

Dennis de Munck
Director of recruitment, L'Oreal

Stéphanie Nguyen
Project leader. L'Oreal

Vasiliki Tseperka
Emotional
Cosmetics
P178

Sheena D. Fraklin
Beauty blogger

Alisa Marie Beyer
Founder. The Beauty Company

Tania Dimitrakopoulou
Korres cosmetics

Harriet Kelsall
Jewellery designer

Renishaw Advanced Materials

AD Baker
Retail design

P288
Nova
Jesse Williams

CRDM
Rapid prototyping

Rich Hartle
Furniture and retail designer

Sensio
Furniture lighting

The Vehicle Glorifier

Jaguar Land Rover Project

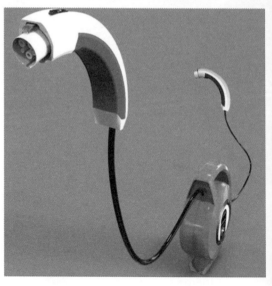

The product is user-friendly. The handle on top of the charger allows the product to be easily grabbed and carried. The grip on the coupler ensures that the user can firmly grab hold of the product. The auto retractable cable system eliminates the need to unwrap and rewrap the cable manually before and after the usage. This reduces the time consumed and on the other hand, makes the charging process fun and pleasurable. The uniqueness of this product is that it is able to stand on the ground and the bright orange colour will make the charger more noticeable. The LED indicator on the back of the charger allows the product to be noticed from every direction and in terms of safety, this can prevent people from tripping over the cable.

K.J Buem, C.C Chuan, S. Kaenratana, J.E Lee

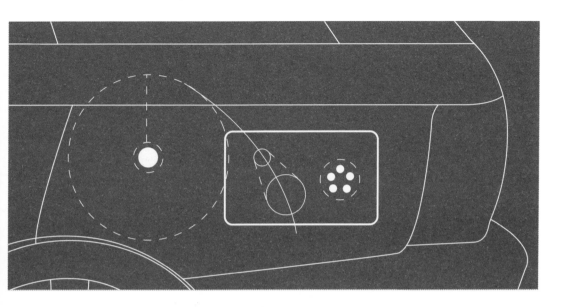

This project arose from an opportunity to work with Jaguar Land Rover to undertake a re-design of the charging cable for the Range Rover HSE Sport. After assessing the market and the customers through content analysis of interviews and questionnaires, it quickly became clear that this project was not about a cable. The problem lies with behaviour and how design can have an impact on changing and improving learnt behaviour. To do this the project was approached from a human centred perspective, looking at how learnt behaviour can be utilised to introduce new concepts and ideas.

The majority of electric vehicles on the market have their plugs and sockets in the front of the vehicle. Standard design is to always put fuel into the back of a car so the system was designed to be mounted into the back wing on the opposite side to the fuel tank. Charging a vehicle can be time consuming and a lot of effort so it was important we took on the task to make the process of charging as effortless and inviting as possible. The cable can be ejected from the console inside the car. The cable is housed in a dock which will eject and by the time the driver has walked to the

back of the car it will be ready for them to take, as the plug is hand held and pulled lightly, the motor on the coil will start again and fully deploy the cable. Once charged, the driver takes the plug from the charge point, a button is pressed and the cable is retracted and docked into the car. The engineering is based on a slip ring which allows the cable to coil without causing fatigue or damage. This system allows for one end of the cable to be permanently attached to the car decreasing the risk of theft.

This system considers the Land Rover customer in many ways. The system is quick and easy to use, it is not messy and can be understood instantly as it uses previously learnt knowledge for its operation. The existing plug design is not adequate for the Land Rover customer or indeed brand. The redesign includes further attention to customisation, colour matching, ergonomics and anthropometrics and quality. Being a premium brand Land Rover must continue this quality through all aspect of the product. Elements of the HSE Sport's design were incorporated into the plug giving the plug not just brand continuity but ownership to Range Rover.

L.B. Amas, E. Bozovali, B. Buist, C. Durucan, S. Kanchi, J. Kim
Design Masters

Eco Car Company

A revolutionary way to power vehicles and apply new technology

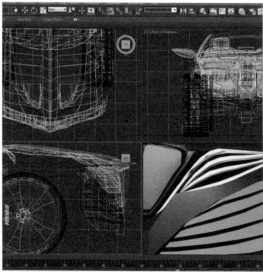

This project explores changing the world by introducing a new way of powering vehicles. This virtual innovation incorporates pioneering new technology, virtually invented by the Eco Car Company; creating a unique concept in the world of vehicle design. It combines overwhelming performance with technology, never before seen in the history of the human race. This is the most intense experience of power and dynamics imaginable in an eco automobile without even having the worry about ever fuelling the vehicle. Unbelievable! The interface provides an interactive experience where the user can explore every aspect of the vehicles. The concept was built with programs such as Unity3d, Flash and many more to create the ultimate driving experience.

Aaron Norman

Multimedia Technology and Design

Influencing Sustainability Through Innovative Design

Encouraging sustainable behaviour in buying and driving

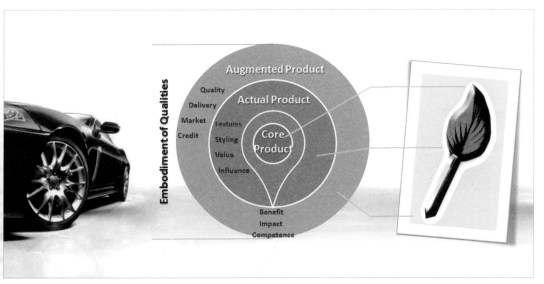

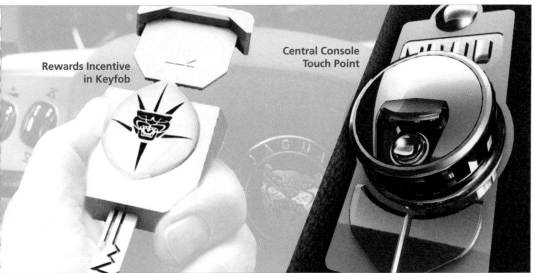

Addressing sustainability is a challenge for everybody. Legislation, environmental targets, governance, economic initiatives and 'big picture thinking' are all valid attempts to shape our future in a more sustainable way. As far as people and businesses are concerned, sustainability comes with extensive cost and time implications, quality compromises and additional running expenses. Current attempts to address sustainable issues within an organisation are frequently seen as only reactions to financial pressures and long term strategic positioning. However sustainability does not have a shape, form, or size. This project investigates how Jaguar could encourage sustainability in people's buying and driving behaviour.

Mohammed Elsouri

Integrated Product Design

Active Suspension Control Testing

Development of test rig

This project involves the development of an active suspension test rig in order to evaluate and improve methods of control. The rig is based on a quarter-car suspension model and is similar to that of a passive suspension system, but with the addition of an actuator which applies a reactive force in response to excitation i.e. bumps and potholes in the case of a real car. The ultimate aim is to improve the road handling and ride comfort of the vehicle.

M. Abukar, A. Arancon, F. Mahmoud, A. Panayiotou, B. Rogunath

Mechanical Engineering

Bicycle CVT and Control System

Prototype continuously variable transmission for a bicycle

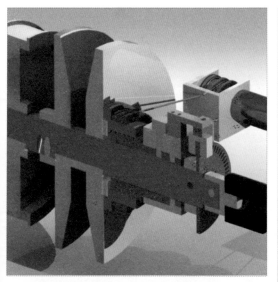

This is a development of a prototype Continuously Variable Transmission (CVT) and automatic computer control system for a bicycle. Using existing CVT technology – the rubber belt CVT – a prototype transmission was developed alongside an automatic, microprocessor based, control system, which when implemented on a bicycle would allow a rider to maintain a constant pedal speed without having to manually change gear. During development, a number of experiments were undertaken to refine the design to optimise mechanical efficiency, FEA analyses were undertaken for weight/strength optimisation, and mechanical design concepts were explored to allow servo control of the transmission and packaging on a bicycle.

Luke Steele
Mechanical Engineering

Better, Slower
Slow Design for a brighter future

Some could argue that ever since the Industrial Revolution, the human race has become preoccupied with doing everything faster. Whether it be faster cars, fast food, or faster internet speed, the impatience of society has never been more pertinent than it is today. At first glance one may think that Slow Design is nothing more than a drawn out temper-tantrum; a reaction from decades of designers becoming tired of working to tight deadlines and having their creativity forced and stifled. It is not about apathy, or a command that everything should be done slower, more an admission that frenzy is not the same as efficiency, and a suggestion of a new work paradigm that may actually help change the world for the better.

In our modern, ruthlessly competitive economic climate, companies are often driven to force products to market in as little time as possible, to keep up with their competitor's new technology, or to avoid being out of fashion. These designs are often poor reflections of the original idea, distorted and warped by the need to produce the end result yesterday. In many cases, the compromises made to fit a tight development cycle are invisible to the consumer, and may only emerge as minor complaints when a product doesn't work the way the user expects or hopes they will, though sometimes the effects are far more serious, occasionally resulting in product failure, injury and product recall.

No one person is to blame for the culture of everything "faster cheaper" that began to emerge most prolifically towards the end of the twentieth century, moreover it can be argued that it is a product of society and economics themselves. It seems that the primary evaluation tool for anything today is economic success, thus the primary goal of any product is to be profitable. But if modern design is the quest to balance economic, social and environmental needs, why are the needs of the user seemingly secondary to the needs of the financier? Money may not be the root of all evil, but it clearly doesn't always help make the right decisions.

The current state of design is intrinsically driven by economic factors, and the primary method of evaluating a design's success is financial. However, as the Microsofts and Sonys of the world will tell you, this contrived focus on economic success is beginning, ironically, to become unprofitable. Those companies slow to adapt to what their customer base actually wants have suffered in an market where purchases are often guided by the collective opinion of the internet. Through these unprecedented channels of communication, products that offer sub-standard user experiences can be publicly ousted, as opinion-based purchasing evolves to a global level.

The term 'Slow Design' was probably first coined by British designer Alastair Fuad-Luke in his 2002 paper "Slow Design - A paradigm for living sustainably". Fuad-Luke posited that despite some progress towards environmental considerations and inclusivity within the profession;

Design [still] hovers in an ideological vacuum and designers wear the mantle of stylists to the powerful.

As Slow Design is still in its infancy as a design philosophy, it is widely open to interpretation. However, most versions of the philosophy share common elements. Taking the moniker of slow design literally, an extended design process does, by its nature, allow more time for a deeper understanding of certain facets of a design, most notably the issues of sustainability and user experience. A slower process intrinsically ushers in elements of long-term thinking, design for longevity and sustainability taking precedence over the cold-hearted financial focus of built in obsolescence.

So the question is; can Slow Design become more than an abstract quest for design perfection? or perhaps more pertinently in today's world, can slow design ever be *profitable*? Closer inspection of the products and services that both the design community and society as a whole have long lauded as 'good design' suggests that perhaps it already is.

Long ago, co-opted by marketers and over-zealous design critics, in many cases 'innovation' is little more than a buzzword or a misguided attempt at positive association. But when we look at products that truly embody the 'spirit' of innovation – the Dysons and the OXO Good Grips of the world, the qualities that make them so are obvious, and are almost always the product of design process whose goal is not constrained by temporal and financial shackles.

Slow design philosophy embodies one of the great paradoxes of design - which is more important, style or substance? In the traditional, economic and marketing led fast-design model, the focus seems firmly shifted towards style or superficial functions, in an effort to keep up with changing fashion trends and competitors. Slow design instead focuses on specific user needs, a true substance-based approach. By creating devices that are exciting and pleasant to use, Dyson, OXO Good grips, the ubiquitous Apple and others like them have kept their focus on the user, and have, somewhat ironically, created designs that transcend fashion; attractive without temporal constraints - true design classics.

So why 'slow' design? If modern design is the effort to balance economic, social and environmental needs, then this new paradigm seeks to remove the economic and therefore temporal constraints from the equation, hence 'slow'. With a greater focus on individual needs, a deeper understanding of sustainability and longevity, and a focus on substance over style or fashion, it is the hope of slow advocates that this new philosophy can save designers from their current positions as organ grinders to the economy and get them back to their rightful place as life-improvers.

Critics of Slow Design dismiss the philosophy as an unrealistic longing for a designer's utopia - a life without deadlines.

How can you remove economic considerations from an industry that is driven by economic success? The simple rebuttal is this; by creating commodities that deeply satisfy the needs of the user, while considering the environmental impact of such commodities, they will naturally attract economic success, as consumers no longer tolerate products and services that do not. In many ways this situation has already arisen - looking at recent economic successes of companies who are seemingly willing to ignore the pressures of the bottom line in pursuit of customer happiness, compared to the economic downturn that stagnant, short term profit-focused companies have seen.

By utilising the power of the Internet and the wider human economy to evolve opinion-based purchasing to a global level, we can create a system of Darwinian consumerism - where the unsatisfactorily weak die off and the well designed strong survive. Slow Design offers us the opportunity to not only save our reputations as designers, but coupled with the evolution of our global economy into a shift away from materialism into 'enoughism', maybe even save the world.

David Elmer

Product Design Engineering

Brunel Racing Formula Student Car

Brunel Racing 2012 car BR-13

Brunel Racing design, build and race their formula student race car among some of the best teams in the world. The team comprises of 36 Engineers from BEng and MEng courses. Together they work on designing a 'lightning-fast' race car, which is driven by members of the team in events at both Silverstone and Hockenheim.

BR12 Design Overview

Brunel Racing is having its hardest year for many years with minimal workshop access for the majority of the year. The strength of the team has really shone through with dedicated team members and helpful sponsors really pulling the project back on track. This will be Brunel Racing's 13th year of competing in Formula Student and the hopes are as high as ever. BR13 is the fourth iteration of the hybrid aluminium honeycomb monocoque/rear spaceframe chassis, which this year incorporates a closed back to further increase chassis stiffness.

Specification
Length / width / height: 2736 / 1460 / 1050 mm
Wheelbase: 1580 mm
Weight of car: 205kg
Front / Rear Track: 1200 / 1175 mm
Weight distribution: 48:52 front to rear
Suspension: double unequal length A-arm pull rod actuated horizontally oriented spring damper
Tyres: 20.0x7.5-13 Hoosier
Wheels: Braid 13-inch Aluminium Rims
Brakes: Grey Iron discs, hub mounted, 220mm dia, drilled, AP Racing four pot calipers front and two pot rear
Chassis construction: Aluminium Honeycomb Monocque Front and Steel Spaceframe Rear
Engine: 2007 Yamaha YZF-R6 Four Stroke inline Four
Bore: 67 mm
Stroke: 42.5 mm
Cylinders: 4 cylinder 599 cc
Fuel system: Bosch Multi Point Fuel Injection
Max power: 60kW @ 11,500rpm
Max torque: 62Nm @ 8,000rpm
Transmission: single 520 chain
Differential: Drexler limited slip differential
Final drive: 3.3:1

Each component, or system of the race car, is designed by a level 3 team member and then collated by a level 5 manager. This design process will include the idea generation and evaluation of previous years' designs. The digital simulation and evaluation using CAD and other software packaging. Then physical testing of the component is often utilised before the final design is constructed. These designs are then manufactured by the students so that the complete car can be constructed ready for pre season testing and the competitions. Working as a team, taking designs through from initial ideas to a car capable of the sub 4 second 0-60, is a truly mammoth achievement.

Brunel Racing Team
Motorsport Engineering

Brunel Racing Formula Student Car

Brunel Racing 2012 car BR-13

Formula Student is a testing ground for the next generation of top engineers. It challenges university students from around the world to design and build a single-seat racing car, which is then put to the test at the famous Silverstone Circuit. The competitions involve the team being tested in their knowledge and understanding of design, cost and technical ability. The car is also put through its paces in the dynamic events which focus purely on the performance of the car and the skill of the driver. As well as the actual competitions, Brunel Racing hold several publicity events as well as attending events such as the Autosport Show and the prestigious Prescott hill climb.

Brunel Racing Team
Motorsport Engineering

Olly Self and Chris Muir
Team Principals
Me07oos@brunel.ac.uk
Me07ckm@brunel.ac.uk

Adam Boix
Chassis Designer
Me07aab4@brunel.ac.uk

James Ackers
Powertrain
Me07jja2@brunel.ac.uk

Mukesh Mehbubani
Vehicle Dynamics & Track Testing Manager
mukesh.mehbubani@gmail.com

Florian Talou
Aerodynamics and bodywork
flotalou@gmail.com

Heather Cray
Unsprung
Me07hhc2@brunel.ac.uk

Richard White
Driver Controls
Me07rrw@brunel.ac.uk

James Thomas
Electrification
Me07jjt3@brunel.ac.uk

Brunel Racing would like to thank DNA Performance Filters, Bosch, Forest Environmental and all our other sponsors, without whom this year would really not have been possible.

Website
www.brunelracing.co.uk

Facebook
www.facebook.com/brunelracing

Twitter
http://twitter.com/brunelracing

Brunel Racing Team
Motorsport Engineering

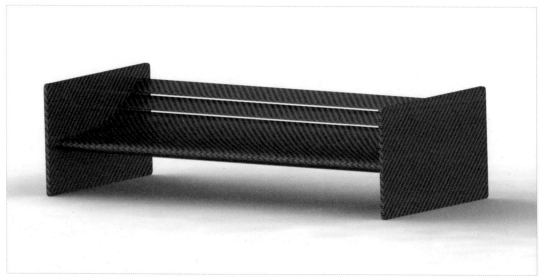

The development of wings is vital for the overall aerodynamic performance of the vehicle. It allows for the production of downforce by taking advantage of Bernoulli's principle. As the air flows over the wing the air molecules move faster over the lower surface of the aerofoil and slower on the upper surface. This creates a pressure difference and a force in the direction of high to low pressure.

The force produced increases the traction between the road and tyres surface for increased stability during, and an allowance for attacking the corners at higher speeds. However the aerodynamic wing itself is not as effective when it comes to drag reduction. Hence the objective was to find an aerodynamic balance between the effective downforce produced and least drag induced.

Protik Sarkar

Motorsport Engineering

The exhaust system has a significant effect on the overall performance of an internal combustion engine. Therefore, careful design, as well as in-depth analysis, is required to optimise an exhaust system for a certain engine. As emission regulations are absent from Formula Student competitions, the system consists of a manifold and silencer.

The new system provides superior performance to the exhaust system from the previous car (BR-12), based on the requirement for improved mid-range performance, supported from track testing data. This has been validated through extensive computer simulation and analysis.

Samuel Robert Moss

Motorsport Engineering

Road Surface Simulation Rig

Active suspension rig for simulating real road profiles

The main aim was to set up a suspension rig and Stewart Platform, to enable a realistic test of a quarter car model; on a simulated road profile. FPGA and CompactRIO were used to control the pneumatic actuators, to provide the excitation to the suspension system and to control the feedback. They have been used in several applications in the past to simulate a road profile for testing suspension systems that are passive, semi-active or fully active. The use of these platforms is also widely seen in biomechanics studies and for simulator studies in other automotive and aerospace applications.

M. Behesthi, F. Caratella, M. Firdaus, J. Gibbon, A. Patel, A. Ruddin, D. Sappal

Mechanical Engineering MEng

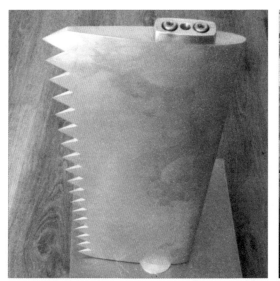
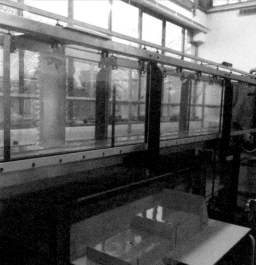
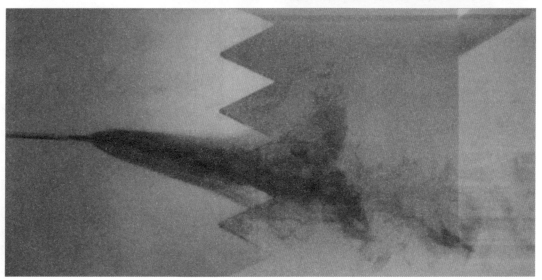

In the aerospace industries, engineers have started designing the trailing edge of the lifting surfaces (e.g. aerofoil) to contain a zig-zag serration pattern for the purpose of noise reduction. A similar chevron pattern is used near the engine exhaust of a BOEING 787. The mechanism of noise reduction by this pattern is mainly attributed to the flow interaction near the serrated edge but its topology remains undefined. This project aims to "see" how the flow interacts near the serrated edge by injecting food dye of different colours using a water tunnel.

Ziaul Rouf
Aerospace Engineering

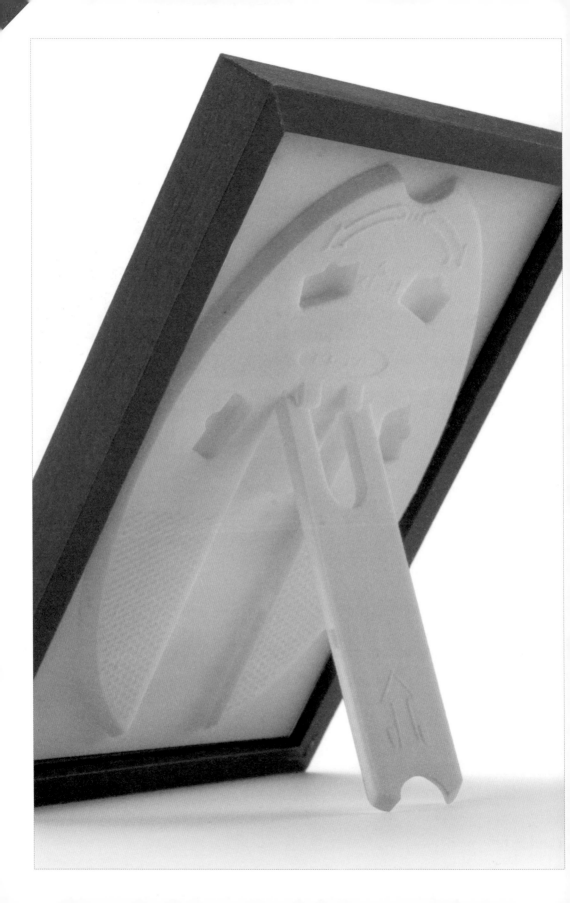

Whether it is ease of use, comfort or aligning your photo, almost everyone has problems with photo frames, but it is a problem many people just put up with. Easy Back, a new attachment mechanism, is an all-in-one injection moulded; retrofittable solution targeted at a number of existing frame styles. Designed to reduce the problems associated with framing your precious photos, Easy Back replaces the awkward mechanisms and catches often used to secure frame backs with an easy to use twist lock. Unlike existing mechanisms, Easy Back locks and unlocks with a simple twist, helps you align your photo inside the frame, or on a mount, and holds the photo securely in place both during insertion and whilst on display.

Cianan O'Dowd

Industrial Design and Technology

RC Camera Dolly

Smooth rotary and linear motion for accurate digital filming

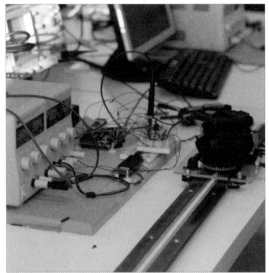

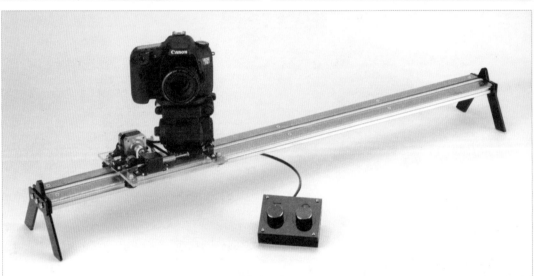

The remote controlled camera dolly provides smooth rotary and linear movement for filming or photography using any compact digital camera. The carriage runs along the rail via belt drive and can rotate the camera 360 degrees in either direction. An innovative variable resistor interface allows for the movement to be controlled via two rotary dials. The motion to be set to any speed or controlled variably to follow a target. Both linear motion and rotation can be set to extremely slow speeds and left running, allowing for time lapse photography. Switches disable the linear motor when the carriage reaches the end of the rail, which can be easily extended when longer shots are required.

Adam Wadey

Product Design Engineering

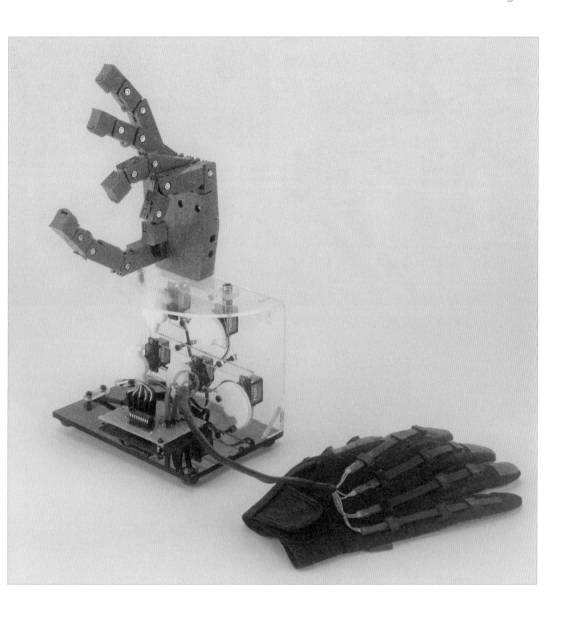

The M.I.M.I.C (or Massively Intuitive Mechanically Integrated Control) Hand gives users ultimate control of a mechanised puppet hand. By wearing a glove with flex sensors on each finger, their movements are tracked and mimicked in real time by a servo controlled puppet hand. The puppet hand is currently mounted on a decorative display stand but has real world applications in fields such as medicine, deep sea exploration, bomb disposal or any task requiring remote control and intricate movement. Utilising knowledge gained from this project a future aim is to develop haptic 3D CAD modelling techniques which could democratise complex design and manufacturing processes.

Edwin Foote
Product Design Engineering

Aperta Industria

Energy harvesting powered door for disabled access

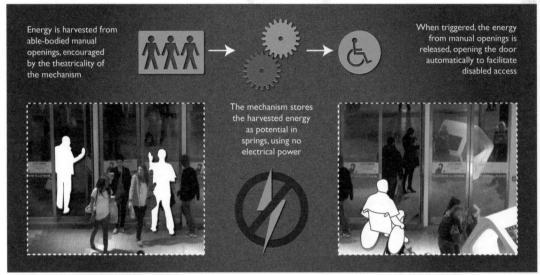

Energy is harvested from able-bodied manual openings, encouraged by the theatricality of the mechanism

The mechanism stores the harvested energy as potential in springs, using no electrical power

When triggered, the energy from manual openings is released, opening the door automatically to facilitate disabled access

Aperta Industria is a conceptual project to design a human powered automatic door for disabled access. By using a fully mechanical system, the device uses zero electrical power - the power for automated opening coming from the energy harvested from manual openings - creating a from of displaced chivalry while removing the electrical demand of conventional powered doors.

An exercise in persuasive design, the system uses a theatrical clockwork-inspired mechanism enclosed within the glazing of the door, encouraging manual opening by making the system interesting and exciting to use, thereby decreasing the chance that the automated functionality is abused.

David Elmer

Product Design Engineering

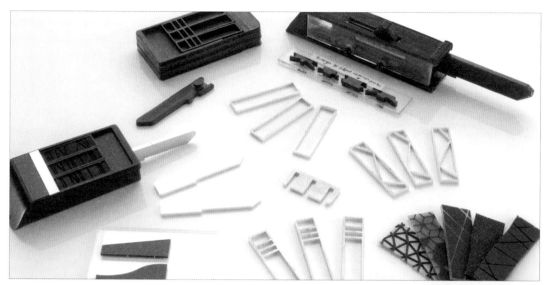

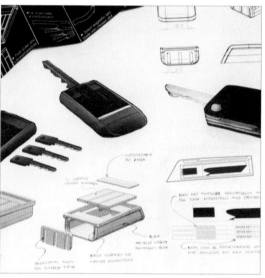

The UK is home to over 26 million households, 4.5 million businesses and there are currently 34 million vehicles on the roads. These forms of ownership are subject to a multitude of daily interactions, facilitated by a commonly disregarded yet fundamental aspect of modern living - the key. Keystone is a key consolidation device, housing a total of three interchangeable and easily accessible keys. Implementation of a spring-assisted mechanism allows for the keys to be both propelled and retracted in an appealing and efficient manner. Keystone aims to enhance the user experience and redefine the perceptions associated with conventional forms of key usage.

Oscar Bowring

Industrial Design and Technology

PVA/Carbon Nanotube Composite Yarn

Design and production for future aerospace applications

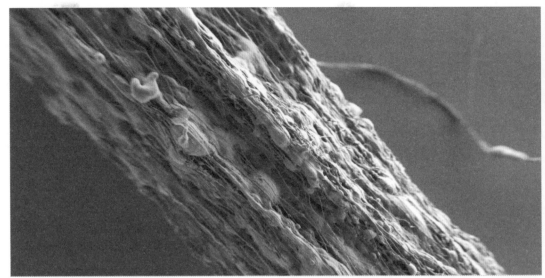

Carbon nanotubes (CNT) are fairly new substances that are recently being studied and investigated in the world of materials which have a promising future in the Aerospace industry if certain challenges are to be overcome. Spinning CNT fibres is an essential step and hence has been invetigated in this project. Composite solutions made of PVA solution and CNT dispersions are mixed together using different concentrations and methods which are further spun with the aid of different spinning techniques. Those include the use of an electrospinning machine, a newly designed spinner and injection of the solution into a diethyl ether bath.

Christina Fraij

Aerospace Engineering

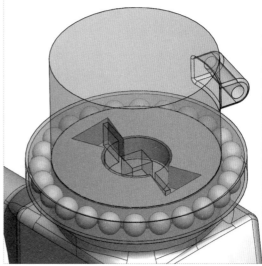

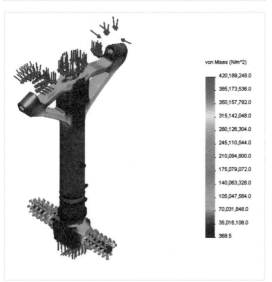

Modelling and analysis of a landing gear system for the A320 aircraft family. One of the main features of the project deals with cross-wind landings. Landing in these conditions can often be a problem, and cause major damage to the gears, resulting in eventual failure. A proposed method aims to reduce the torsional loading on the landing gear during cross-wind landings, by implementing a system that will line the wheels up with the direction of the landing strip, and gradually align the aircraft, once landed, for minimum torsional stress. This system will aid landings and ensure the gears to not experience forces greater than those to potentially cause damage, and will also improve current safety of these dangerous landing manoeuvres.

R. Dhawan, A. Matthews, P. Sarkar, U. Taqi
Aerospace Engineering

Squid Grip

Biomimetic boat hull cleaner

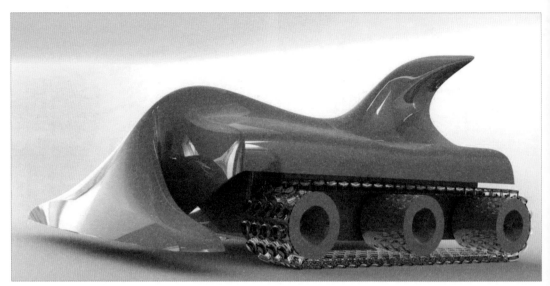

Squid Grip Map
A Design Journey

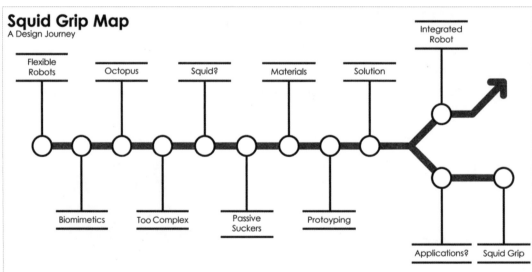

Flexible Robots — Octopus — Squid? — Materials — Solution — Integrated Robot

Biomimetics — Too Complex — Passive Suckers — Protoyping

Applications? — Squid Grip

Biomimetics is a growing subject area worldwide. Research in this area often too often results in interesting technology that is rarely used in marketable applications. Squid Grip is an example of how product designers can help bridge the gap between interesting technology and integrated product. Squid-inspired suckers enable Squid Grip to stick to the hull of the boat. The device navigates itself using sensors and a water blade is used to remove algae. Regular use means that small boat owners can keep their hulls clean without having to remove their boats from the water to carry out maintenance.

Tim Palmer Fry
Integrated Product Design

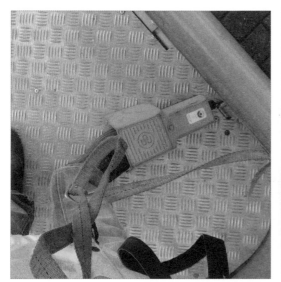
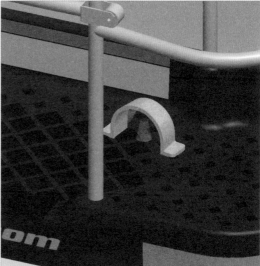

Mobile Elevated Work Platforms (MEWP) are used in a variety of areas in which access at heights are required and have been increasingly common in industrial projects. Operation of these machines requires a switch that has to be pressed and held down to activate the control panel. The most common type comes in the form of a foot pedal. The Niftylift foot device looks into improving the ergonomics and ease of use of a hold-to-run switch, whilst mitigating against misuse.

Jacob Lee
Product Design Engineering

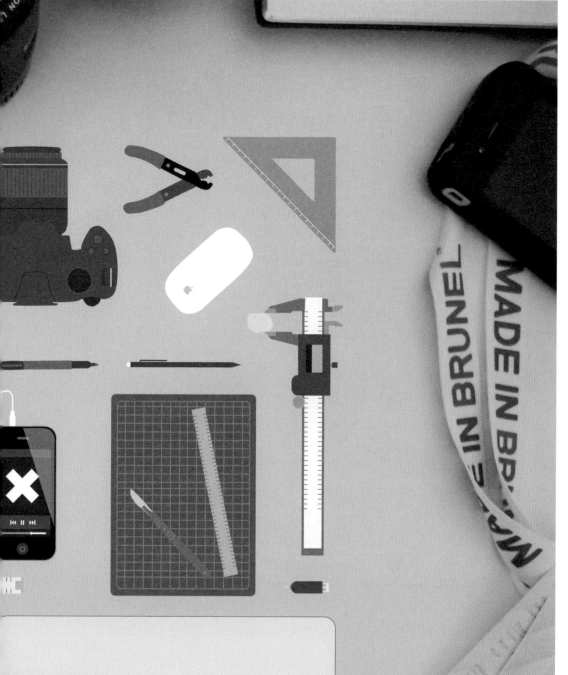

MADE IN BRUNEL

journeys
fuelled by ideas
14th - 17th June

Automated PCB Surface Cleaner

Automatic contamination removal mechanism

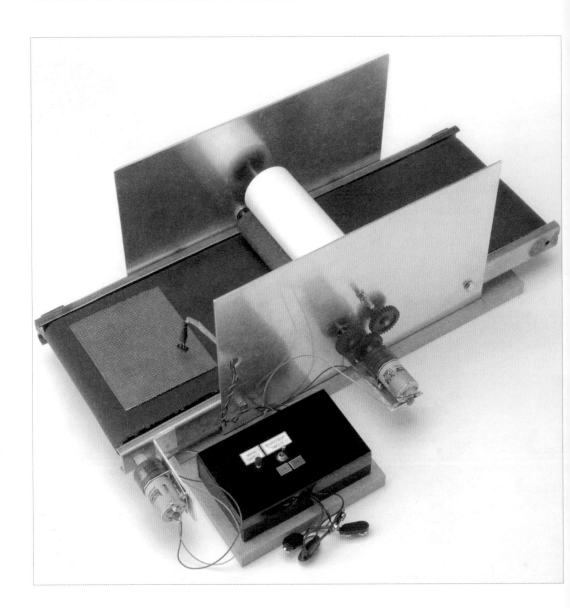

An automatic contamination removal mechanism designed to optimise the cycle time and work operation as well as to eliminate manual working procedures. Dust particles on the surface of the PCB board could inflict damage to the circuit and thus increase the quantity of rejected parts. At present, dust is removed manually by applying a roller with a sticker on the surface of the PCB.

The main components in this product to enable top surface cleaning are a conveyor belt and two types of rollers, which will be in contact with the PCB surface. Dust particles from the PCB material are being transferred to the surface of the elastomer rollers, and followed onto the surface of the adhesive roller.

Shazwani Farhah Shamsuddin

Product Design

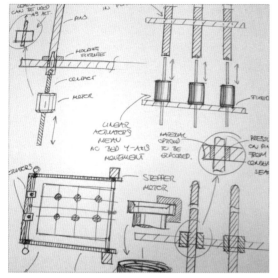

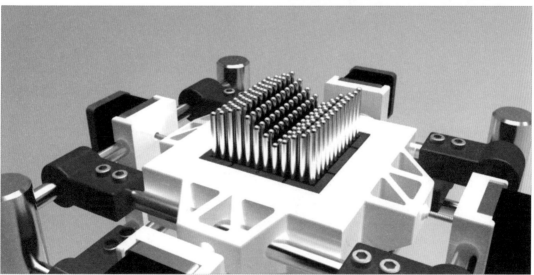

Advances in rapid prototyping technologies are making product development processes increasingly efficient, yet the process of creating and modifying vacuum forming moulds is still very costly and time-consuming. This design offers a means for designers and manufacturers to mould any design in one reconfigurable tool in just a matter of minutes, which can be configured directly from 3D CAD data or by manually pushing existing parts into the array of 2mm diameter discrete pins. The system makes creating moulds and implementing modifications flexible, cost-effective and substantially quicker. A product architecture has been developed that provides a system which can be scaled according to the required working aperture.

Patrick Bion
Product Design Engineering

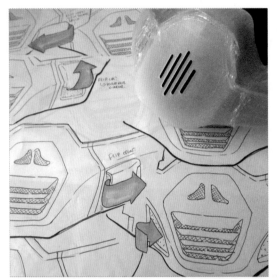
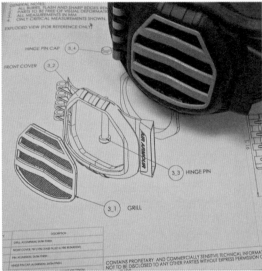

One in five veterans from Afghanistan and Iraq have been shown to suffer some form of Traumatic Brain Injury (TBI) from repeated exposure to Improvised Explosive Device (IED) blasts. Air Armour provides protection against TBI by deflecting this blast wave ensuring it does not enter the brain. Design development focused on improving the comfort of the device whilst meeting robust technical requirements. Air Armour consists of a lightweight composite shell, a shock absorbing mechanism and flexible rubber skin that helps to protect against blunt trauma forces. Respiratory protection is also incorporated and a supply of cool blown air over the face helps to reduce heat fatigue, a common with other protective equipment.

James Carswell
Industrial Design and Technology

in2 Lounge Chair
Improving storage, comfort, support and efficiency

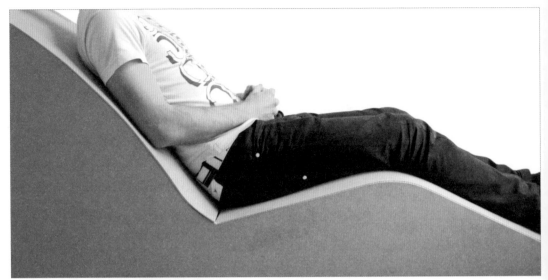

The majority of contemporary lounge chairs require large amounts of space and are often difficult to place in storage; this design proposal is intended to provide a solution by allowing a single disassembly operation. The lounge chair is unique in its contemporary design with two identical components creating its form. These components are easily stackable and reduce storage space for both consumer and retailer. Design for manufacture has influenced the development of the in2 lounge chair, unlike many of its competitors. Material properties and performance are integral elements of the project, ensuring that the visual qualities are not compromised. The lounge chair has a strong focus in the support and comfort provided to the user.

Jack Sandys

Industrial Design and Technology

Safety Footwear for UHP Water Jetting

Ergonomic protection from extreme point impact

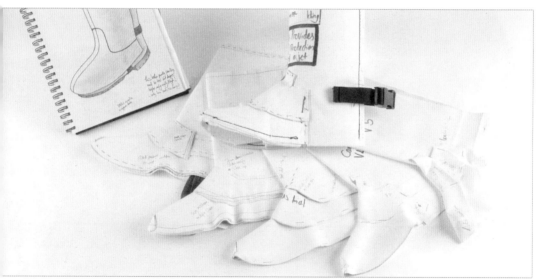

Currently available safety footwear for ultra-high pressure (UHP) water jetting is very poorly designed. Protecting the wearer's feet and lower leg from jet pressures of up to 3,000 bar (43,500 psi) current protection is heavy, inarticulate and causes extreme fatigue to the wearer. As such, many jet operators opt to go without wearing such protection, leaving themselves vulnerable to severe, and even life threatening, injury. The final design is an ergonomically comfortable, lightweight and flexible alternative using a composite of advanced materials. The final product is more flexible and lightweight than current models and have the additional advantage of being almost half the cost to produce of the competition.

Joe Carling

Industrial Design and Technology

Nova

Value lies in the creativity

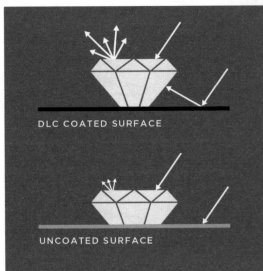

DLC COATED SURFACE

UNCOATED SURFACE

BLACK
— is the new —
BLACK

As new materials have been discovered, they have been investigated to develop new forms and functions. The design and manufacture of jewellery as an exploration tool is an example of the way in which materials, not necessarily precious, can be made valuable because of novel properties. The project explores this concept with a new material combination - Diamond-like Carbon coatings, and cut diamonds. The investigation has involved a deeper understanding of emotive design and the utilisation of a design process to create an appealing product. The objective was to create a user experience which adds value to both the product itself and the perception of its quality. This is achieved through a designed collection of jewellery and a method of experiencing it.

Jesse Williams

Industrial Design and Technology

Our Journey

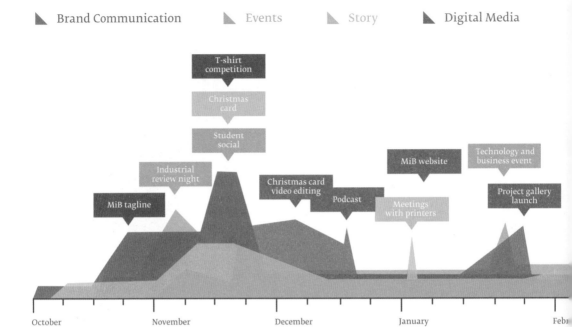

Brand Communication   Events   Story   Digital Media

T-shirt competition

Christmas card

Student social

Industrial review night

MiB tagline

Christmas card video editing

Podcast

MiB website

Meetings with printers

Technology and business event

Project gallery launch

October   November   December   January   Febr

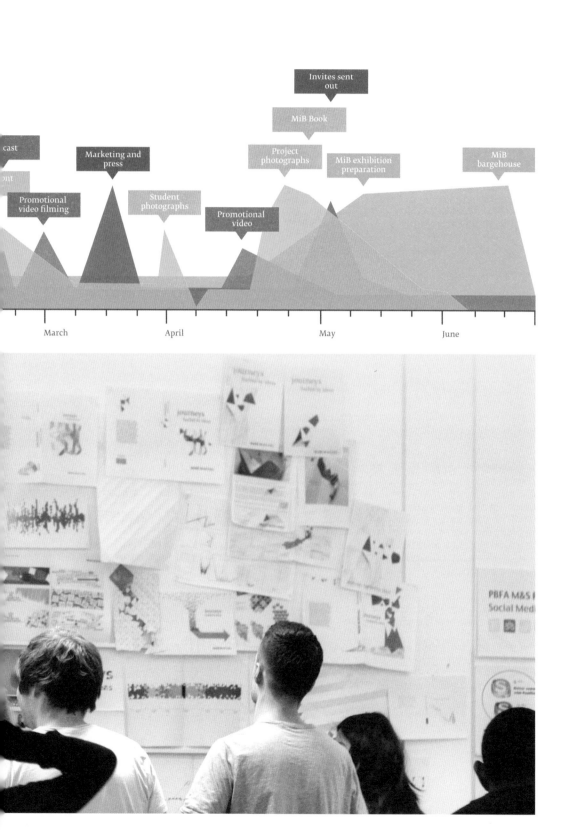

Invites sent out

MiB Book

cast

ont

Marketing and press

Project photographs

MiB exhibition preparation

MiB bargehouse

Promotional video filming

Student photographs

Promotional video

March   April   May   June

PBFA M&S
Social Med

Brunel University

The history and ethos of Brunel University

Brunel is a world-class university based in Uxbridge, West London. Our distinctive approach has always been to combine academic rigour with the practical, entrepreneurial and imaginative approach pioneered by our namesake, Isambard Kingdom Brunel.

The University was founded in 1966 as a new kind of institution dedicated to providing research and teaching that could be applied to the needs of industry and society. This goal was central in the creation of our Royal Charter which emphasised the University's commitment to the relevance of academic learning, ensuring that teaching and research benefit both individuals and society at large. The confidence and sense of purpose that characterise today's Brunel stem from this principle, which is still central to our Mission, "to advance knowledge and understanding and provide society with confident, talented and versatile graduates", and Vision, "to be a world-class creative community that is inspired to work, think and learn together to meet the challenges of the future".

Research is at the heart of all we do, not just the preserve of academic staff. All Brunel students complete a final-year project built on research and the application of knowledge. We place great value on the usefulness of our research, which improves our understanding of the world around us, informs up-to-the-minute teaching, as well as creating opportunities for collaborative work with business, industry and the public sector. Moreover, Brunel's research ethos generates an atmosphere of innovation which inspires students and staff throughout the University, as well as encouraging the sharing and communication of ideas and expertise for which we are famous.

Brunel's reputation is built on its groundbreaking work in subjects as diverse as design, engineering, education, science, sociology, IT, psychology, law and business. That reputation has gradually spread to new areas, including the performing arts, journalism, environmental science, sport and health; however research and teaching in these areas still retains the same outward-looking philosophy.

Brunel's journey through a long succession of developments and mergers, particularly those which transformed Acton Technical College into Brunel University and which later saw Brunel merge with Shoreditch College and the West London Institute, have made the University a major force within the UK higher education sector and on the international stage. Brunel's campus has been transformed over the last few years now boasting world-class sports facilities, more green and social space for staff and students, renovated laboratories and classrooms, as well as an extended library with a hugely increased book and journal collection, more computer workstations and group study areas and an Assistive Technology Centre for disabled students. The Schools of Health Science and Social Care and Engineering and Design have benefitted from new buildings and facilities and a flagship building at the main entrance to the University has recently been completed, providing a new home for the Brunel Business School as well as an auditorium, conferencing facilities and an art gallery.

Since the 1960s, Brunel University has provided high quality academic programmes which meet the needs of the real world and contribute in a practical way to progress in all walks of life. We are very proud of the University's development and of its contributions to the empowerment of individuals and to the progress of society. We have evolved a culture of innovative thinking that is applied, not just ideas in isolation but realised ideas. We apply our creativity and ingenuity across the broadest spectrum of subjects, working to add value and to build viable and exciting solutions to the challenges of the modern world. We are future thinkers.

At the heart of Brunel, the School of Engineering and Design continues to draw inspiration from Isambard Kingdom Brunel, widely recognised as one of the UK's finest engineers and also as embodying the principles of design thinking: applying human centred creative approaches to solve complex challenges.

Design, engineering and innovation will be crucial to responding to the challenges facing us today. The School is globally recognised to be a centre of excellence in these vital disciplines with an enviable reputation as an innovative provider of educational and research programmes which lead to advances in knowledge and skills for sustainable technological development alongside substantial engagement with industry.

The School offers 20 undergraduate courses and 20 postgraduate courses catering to a student body of just over 3000. 320 PhD students from around the world work with leading researchers, 95% of whom are ranked at International standard, involved with multi-million pound research programmes and collaborations. Dedicated technicians, extensive workshop and laboratory resources underpin teaching and research, forming the human and physical resources that create the incredible work within this Book.

Students in the School have a significant head start in an uncertain job market, with the vast majority benefiting from taking a year long work placement in industry. The ongoing commercial relevance of our courses is part of the distinctiveness of our approach and is part of the reason behind the high employment performance of our graduates.

Design

Home to the highest rated design courses in the UK, with 500 undergraduate and over 200 postgraduate students from around the world, Brunel excels in technically and human orientated design. Distinctive qualities of our work are extensive links with industry partners and strong emphasis on the underlying practical skills and knowledge required for work in emerging professional contexts. With over 75% of our undergraduate students completing a year working in industry during their course, this leads to an incredible record of employment in all areas of the design profession. Important specialisms within our work include: environmental sustainability, branding, design management, inclusive design, design for manufacture and CAD.

Electronic and Computer Engineering

Preparing graduates for roles in today's digitally connected society, this subject area boasts strong industry links, working with Microsoft, Avid, Xerox, Dare Digital and many others. The subject area currently manages research projects with a value of over £5 million based around telecommunications and systems. 600 undergraduate and 300 postgraduate students are supported by almost 50 academic staff. Subject specialisms include: broadcast media design and technology, computer systems, internet engineering, sustainable power and electronic and electrical engineering.

Mechanical Engineering

Many of the big challenges facing the world require an engineering component. In the field of sustainability the mechanical engineering subject area has had high profile success such as being the most successful university entry in the electric motorcycle TTXGP event on the Isle of Man, and the 'Water Cycle' human powered water purification device. In addition to this core mechanical engineering expertise, specialisms include aerospace, aviation, building services, automotive and motorsports. Effective team work to integrate mechanical engineering within larger projects is encouraged through team challenges including the highly successful Brunel Racing team.

Advance Manufacturing and Enterprise Engineering

Focusing on postgraduate education and research this subject area develops world class expertise in advanced manufacturing technology, enterprise engineering and engineering management. The AMEE research laboratory has specialist facilities supporting: micro/nano manufacturing/metrology, design of ultra-precision machines, robotics and manufacturing automation, 2D/3D vibration assisted machining and advanced manufacturing using lasers.

Civil Engineering

Meeting the important needs of the 21st Century, especially in sustainable building construction, roads, bridges, tunnels, flood protection, waste recycling and construction management. Sustainability is at the heart of our ethos, paralleling the 'cradle-to-grave' approach promoted by major national and international Engineering organisations. Our technical facility, the Joseph Bazalgette Laboratories, includes the latest technology to develop and test innovative materials.

Isambard Kingdom Brunel and the London Airport Problem

Stephen Green, Brunel University

Shortly after this book is published in the summer of 2012, London will experience an unprecedented influx of visitors for the London Olympic and Paralympic Games. We do not yet know how London's airports will cope, but we do know that they will be severely tested and that the pattern is set to continue with estimated increases in global air traffic of 100% by 2030; yet Heathrow is already running at 98% capacity. What would Isambard Kingdom Brunel do? A rhetorical question which defies objective answers perhaps, but we can consider it in relation to the historical facts of Brunel's methods and achievements. In particular we can consider – as Tim Brown, CEO of IDEO, pointed out – that Brunel adopted a design thinking approach to his greatest accomplishments, before the term had even been invented.

Many of Brunel's greatest achievements are linked to the Great Western Railway, for which he became Chief Engineer at the age of 27, only three years after the world's first passenger railway opened. The challenge to build a rail network at this time could be compared to flying men to the moon when controllable rockets had only just been invented. His appointment was marked by him challenging the basic premise upon which the process was based: arguing that the criteria of building the cheapest route to London for the new railway was wrong. He would offer the best route. This initial line, from London to Bristol is still in use and remains one of the straightest, fastest, routes in the UK 175 years later.

Brunel is often remembered for his 7ft 'broad gauge' spacing of the rails. We can view this as a human centred design approach: his vision was based on delivering comfort and speed, with the specific aim of achieving 80mph versus the then top speed of 40mph. Looking beyond linking cities within the UK, his vision grew to encompass the notion of integrated transport and seamless journeys between London and New York. This was realised with the launch of his Great Western in 1837, the first transatlantic steamship in the world, followed by the Great Britain in 1843, the first significant ship with an iron hull and screw propellers. Brunel's final and tragic ship was the Great Eastern launched in 1859, days before his premature death. This ship remained the largest in the world for 40 years and had been envisioned to carry passengers and all the coal needed for a round trip to Australia. But here Brunel's design thinking touch had perhaps deserted him and there was insufficient passenger demand for the service.

By virtually any measure Brunel's achievements are immense, but, in his prime, his ambition, ability to sell his vision and then make vision a reality through creativity, sound design engineering, planning, leadership and collaborative working demonstrate an unsurpassed quality of design thinking which should be a benchmark for us all. The way he conceptualised the answers to great challenges as innovative systems with human centred concerns at their heart is simply inspirational. We can only hope that some of these qualities will be evident in plans for the future of air travel to and from London.

The quadrant of London sweeping out from Kensington and Chelsea, encompassing Wembley Stadium, Heathrow Airport and the Thames, M4 and M40 corridor travelling westwards is one of the most economically successful and vibrant regions in the world. West London comprises of more than 750,000 jobs and 67,000 businesses, which annually contribute over £27 billion to the UK economy. Heathrow, the worlds most successful airport, is a natural global hub for the world. The region has a tremendous track record as a home for innovative business at the cutting edge of global markets including the likes of GSK, BA, Disney, Coca Cola, Diagio, and BSkyB.

Joint research conducted in 2007/8 amongst large corporates and SME's in West London and funded by the London Development Agency, clearly showed the high level of innovation achievement within West London whilst also identifying a keen appetite for sharing and gaining more knowledge on emerging innovation methods and techniques.

Collaborations between business and universities are increasingly recognised as important catalysts for innovation and economic regeneration. Many of the people and projects which make up Made in Brunel exemplify collaborations to harness the energy, creativity, research and knowledge available within the entrepreneurial spirit of the area. The following organisations and initiatives are part of the infrastructure within West London to provide practical support to facilitate connections and the innovation which can flow from these:

West London Partnership

Chief executives and leaders of the six local authorities in West London and senior staff both from large corporate businesses and from SMEs based in West London join forces as WLP to develop the overall social and economic interests of the region. This dynamic partnership between business and local authorities has a broad remit including transport, regeneration, skills and workforce development, spatial development, planning and property, and housing.

West London Business

A business membership organisation for the region; West London Business promotes economic development, entrepreneurial activity and innovation, as well as supporting the many growth sectors in West London, such as pharmaceuticals, food, IT, logistics, creative industries and tourism. West London Business is committed to raising economic competitiveness across the region, creating new jobs and retaining businesses in the area, while promoting the principles of social inclusiveness and sustainability.
www.westlondon.com

Designplus

Designplus promotes design based collaborations between industry and universities. Based in Brunel University since 2004 Designplus establishes collaborative projects, leads events and provides professional development.
www.designplus.org.uk

Network Directory

Mohamed Abukar

Mechanical Engineering MEng
m.abukar@hotmail.co.uk
Pages: 256

Craig Aburrow

Broadcast Media Design and
Technology BSc
craigaburrow@yahoo.co.uk
www.craigaburrow.com
Pages: 36

Lide Brito Amas

Design and Branding Strategy MA
lide_rizos@hotmail.com
Pages: 253

Anthony Arancon

Mechanical Engineering MEng
Pages: 256

Vincent Ashikordi

Product Design Engineering BSc
vashikordi@gmail.com
www.coroflot.com/
vincentashikordi
Pages: 16, 227, 238

Sukhi Assee

Industrial Design and
Technology BA
sukhiassee@hotmail.co.uk
Pages: 86, 189

John Aston

Industrial Design and
Technology BA
johnaston17@hotmail.co.uk
Pages: 66, 190

Luke Bacon

Product Design BSc
bacon.lukejames@gmail.com
Pages: 88, 223

Georgiana Bădălîcă-Petrescu

Design and Branding Strategy MA
badalicag@yahoo.com
Pages: 176, 177

Matt Baldwin

Industrial Design and
Technology BA
mattb_52@hotmail.co.uk
Pages: 81

Matthew Bastow

Industrial Design and
Technology BA
matthew_bastow@hotmail.com
Pages: 44, 224

Milad Beheshti

Mechanical Engineering MEng
miladbeheshti25@gmail.com
Pages: 266

Bedir Bekar

Civil Engineering with
Sustainability MEng
bedirbekar@gmail.com
Pages: 30

Salvatore Bella

Industrial Design and
Technology BA
salvatorebelladesign@gmail.com
Pages: 117, 233

Patrick Bion

Product Design Engineering BSc
patrick@patrickbion.com
www.patrickbion.com
Pages: 148, 215, 283

Pavlina Blahova

Multimedia Technology and
Design BSc
pavlinablah@gmail.com
www.pavlinablahova.com
Pages: 144

Oscar Bowring

Industrial Design and
Technology BA
oscarbowring@me.com
Pages: 194, 274

Elif Bozovali

Design and Branding Strategy MA
elifbozovali@gmail.com
Pages: 253

Stuart Brockwell

Multimedia Technology and
Design BSc
stueybrock@gmail.com
Pages: 154

James Bruck

Multimedia Technology and
Design BSc
jamesmbruck@gmail.com
Pages: 146

Benjamin Buist

Design and Branding Strategy MA
bbuist1@gmail.com
Pages: 180, 253

Hayley Rose Burrows

Product Design BSc
hayley.rose.burrows@gmail.com
www.hayleyroseburrows.com
Pages: 10, 197

Fazal Caratella

Mechanical Engineering MEng
caratella@hotmail.co.uk
Pages: 266

Joe Carling

Industrial Design and
Technology BA
joseph.carling7@gmail.com
Pages: 53, 208, 287

Sofia Carobbio

Design Strategy and Innovation MA
sofia.carobbio@gmail.com
Pages: 181

James Carswell

Industrial Design and
Technology BA
jamescarswell.id@gmail.com
Pages: 203, 284

Richard Carter

Industrial Design and
Technology BA
rpcarter@me.com
www.linkedin.com/in/rpcarter
Pages: 21

Laura Castiglione

Multimedia Technology and
Design BSc
laura_castiglione@hotmail.com
Pages: 49, 158

Balraj Chana

Multimedia Technology and
Design BSc
balraj.chana@gmail.com
Pages: 159

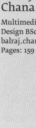

Tajinder Cheema

Industrial Design and
Technology BA
topcatglobal@gmail.com
Pages: 237

Dan Cherry

Product Design BSc
info@dancherrydesign.co.uk
www.dancherrydesign.co.uk
Pages: 84, 186

Peter Cheung

Industrial Design and
Technology BA
ph_cheung@hotmail.co.uk
Pages: 79, 229

Shu Yang Chou

Integrated Product Design MSc
shuyang0129@hotmail.com
Pages: 15, 17

Chong Chin Chuan

Design and Branding Strategy MA
Pages: 252

Tom Cody

Product Design Engineering BSc
me@codytom.com
Pages: 195, 244, 245

Carla Sabrina Conte

Integrated Product Design MSc
csconte@gmail.com
Pages: 15, 18

George Coombes

Product Design Engineering BSc
gwtcoombes@googlemail.com
Pages: 125

Oscar Correa

Industrial Design and
Technology BA
Pages: 231

Ben Crichton

Industrial Design and
Technology BA
bravocharliedesign@gmail.com
www.bravocharlie.net
Pages: 118, 220

David Crittenden

Industrial Design and
Technology BA
dcrittenden.design@gmail.com
Pages: 92, 186

Andrew Davies

Product Design BSc
andrewdavies27@hotmail.com
Pages: 134, 137, 187

Jaymini Desai

Industrial Design and
Technology BA
desaijaymini@gmail.com
Pages: 46, 227

Exequiel Di Salvo

Industrial Design and
Technology BA
exe.disalvo@gmail.com
Pages: 75, 229

Rob Dowling

Industrial Design and
Technology BA
rdowling90@gmail.com
Pages: 116

Yunqian Du

Industrial Design and
Technology BA
du_yunqian@yahoo.com.sgW
Pages: 76, 236

Cansu Durucan

Design and Branding Strategy MA
cansudurucan@gmail.com
Pages: 253

Steve Dyson

Integrated Product Design MSc
s.j.b.dyson@gmail.com
Pages: 13

Samuel Edwards

Product Design BSc
samueldavidedwards
@googlemail.com
Pages: 114, 213

Maryam Abdul Elahi

Industrial Design and
Technology BA
maryam.ae@hotmail.co.uk
Pages: 237

Clayton Jack Ellacott

Industrial Design and
Technology BA
claytonellacott@gmail.com
Pages: 124

David Elmer

Product Design Engineering BSc
davidelmer@me.com
www.davidelmer.co.uk
Pages: 200, 249, 258, 272

Mohammed Elsouri

Integrated Product Design MSc
elsourim1@hotmail.com
Pages: 255

Aaron Faber

Multimedia Technology and
Design BSc
aaron.jack.faber@gmail.com
www.aaronfaber.co.uk
Pages: 158, 161

Mohd Firdaus

Mechanical Engineering MEng
Pages: 266

Jordan Fisher

Multimedia Technology and
Design BSc
jordan.fisher@live.com
Pages: 42

Edwin Foote

Product Design Engineering BSc
edwinfoote@gmail.com
Pages: 50, 199, 271

Christina Fraij

Aerospace Engineering BEng
cfraij91@gmail.com
Pages: 276

Louis Garner

Product Design Engineering BSc
louisgarner349@hotmail.com
Pages: 113, 126, 203

Mitch Gebbie

Product Design Engineering BSc
Mitchgebb@gmail.com
www.mitchgebbie.moonfruit.com
Pages: 14, 136

James Gibbon

Mechanical Engineering MEng
Pages: 266

Victoria
Gibson-Robinson

Product Design BSc
v.gibsonrobinson@gmail.com
Pages: 112, 216

Laura
Ginn

Product Design Engineering BSc
laura.ginn@hotmail.co.uk
www.lauraginn.co.uk
Pages: 40, 224

Luke
Gray

Industrial Design and
Technology BA
this.is.luke.gray@gmail.com
Pages: 60, 131, 207

Andy
Green

Multimedia Technology and
Design BSc
hello@andygrn.co.uk
www.andygrn.co.uk
Pages: 141

Rachael
Greene

Broadcast Media Design and
Technology BSc
ragreene01@gmail.com
Pages: 47

Adam
Greenland

Civil Engineering with
Sustainability MEng
adam.p.greenland
@googlemail.com
Pages: 30

James
Hellard

Industrial Design and
Technology BA
jhellard@live.com
Pages: 132, 197

Cameron
Henderson

Industrial Design and
Technology BA
c.r.henderson16@gmail.com
Pages: 71, 211

Russell
Hinton

Multimedia Technology and
Design BSc
hintonmedia@gmail.com
www.hintonmedia.com
Pages: 130

Laura
Hodges

Product Design BSc
laura.p.hodges@gmail.com
Pages: 102, 204

Benny Ho
Hon Chi

Industrial Design and
Technology BA
Pages: 43

Barnaby
Hunter

Product Design BSc
barnaby.hunter@googlemail.com
Pages: 119, 194

Sarah
Hutley

Industrial Design and
Technology BA
sarahhutley@gmail.com
Pages: 54, 70, 205

Jonathan
Hyslop

Product Design Engineering BSc
jonnyhyslop@googlemail.com
Pages: 93

Alec
James

Product Design BSc
alecj89@hotmail.com
Pages: 90, 197

David
Johnston

Industrial Design and
Technology BA
davidkeiffer.johnston
@googlemail.com
Pages: 108, 212

Hardeep S.
Kalsi

Aerospace Engineering BEng
hardeepis@gmail.com
Pages: 106

Sasikishan
Kanchi

Design Strategy and Innovation MA
ksasikishan@gmail.com
Pages: 253

Mohammad Javad
Karimianzadeh

Civil Engineering with
Sustainability BEng
moejavad@ymail.com
Pages: 29

Stefani
Karaoli

Product Design BSc
stefi_k21@hotmail.com
Pages: 236

Luke Kavanagh

Mechanical Engineering M Eng
kavanaghginge@hotmail.co.uk
Pages: 104

Jung Kim

Design and Branding Strategy MA
evelyn1022k@gmail.com
Pages: 253

Ben King

Industrial Design and
Technology BA
dt08bdk@gmail.com
Pages: 211, 242

Alyssandra Lagopoulou

Industrial Design and
Technology BA
alys_lag@hotmail.com
Pages: 190

Boyeun Lee

Design and Branding Strategy MA
boyeun.lee@gmail.com
http://lby7190.blogspot.co.uk
Pages: 28, 33

Jacob Lee

Product Design Engineering BSc
Jacob_lee89@yahoo.co.uk
Pages: 230, 279

Jeff L.K. Lee

Multimedia Technology and
Design BSc
jeffleeoo777@hotmail.co.uk
Pages: 145

Polina Liarostathi

Integrated Product Design MSc
pliarostathi@gmail.com
www.designpauline.com
Pages: 32, 95, 168

Yonghun Lim

Integrated Product Design MSc
yonghunlim83@gmail.com
Pages: 15, 33, 169

Tim Logg

Industrial Design and
Technology BA
loggtim@gmail.com
Pages: 37, 103, 207

Angela Luk

Product Design Engineering BSc
angela.luk90@gmail.com
Pages: 64, 136

Chao Luo

Integrated Product Design MSc
luochao0330@live.cn
Pages: 167

Christopher Luscombe

Mechanical Engineering BEng
badmintonluscombe
@hotmail.co.uk
Pages: 122

Ferid Mahmoud

Mechanical Engineering MEng
feridmah@googlemail.com
Pages: 256

Alkesh Makwana

Multimedia Technology and
Design BSc
alkeshmakwana@hotmail.com
www.digitalki.co.uk
Pages: 25

Xander Marritt

Multimedia Technology and
Design BSc
xandermarritt@gmail.com
www.xanderjames.co.uk
Pages: 138

Andrew Matthews

Aerospace Engineering BEng
andrew.matthews@hotmail.co.uk
Pages: 246, 277

Simon McNamee

Product Design Engineering BSc
simonmcnamee@gmail.com
www.simonmcnamee.com
Pages: 120, 200

Emily Menzies

Product Design BSc
emily_menzies@hotmail.co.uk
www.emilymenzies.co.uk
Pages: 123, 217, 248

Carolina Montenegro

Design and Branding Strategy MA
carolinhanm@gmail.com
www.flavors.me/carolmontenegro
Pages: 174, 175

Samuel Robert Moss

Motorsport Engineering MEng
meo8srm@googlemail.com
Pages: 265

Cilas do Nascimento Sousa

Design and Branding Strategy MA
cinaso@gmail.com
www.flavors.me/cilas
Pages: 157, 174

Chris Naylor

Product Design BSc
naylor.chrisj@gmail.com
www.coroflot.com/chrisnaylor
Pages: 77, 199

Mina Nishimura

Multimedia Technology and
Design BSc
mina.mishimu@gmail.com
www.mishimu.com
Pages: 152, 158

Aaron Norman

Multimedia Technology and
Design BSc
the_2quick_mixtapes
@hotmail.co.uk
www.aaron-norman.com
Pages: 254

Cianan O'Dowd

Industrial Design and
Technology BA
cianan27@gmail.com
www.uk.linkedin.com
/in/ciananodowd
Pages: 78, 219, 268

Sophie O'Kelly

Product Design BSc
sophie.a.okelly@gmail.com
Pages: 45, 164, 204

Johanna Elise Oja

Multimedia Technology and
Design BSc
johannaoja@gmail.com
Pages: 150, 158

Kingsley Okyere

Multimedia Technology and
Design BSc
kkbokyere@gmail.com
www.kingsleyokyere.co.uk
Pages: 140, 158

David Paliwoda

Broadcast Media Design and
Technology BSc
www.davidpaliwoda.com
davidpaliwoda@gmail.com
Pages: 142

Tim
Palmer Fry

Integrated Product Design MSc
timpalmerfry@hotmail.co.uk
Pages: 19, 278

Andreas
Panayiotou

Mechanical Engineering BEng
andrewpany2002@gmail.com
Pages: 256

Ajay
Patel

Mechanical Engineering MEng
ajay_patel2k@hotmail.com
Pages: 266W

Dhanish
Patel

Industrial Design and
Technology BA
dhanishpatel7@gmail.com
Pages: 37, 189, 240

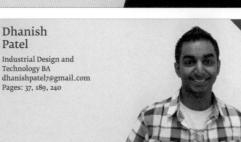

Nejal
Patel

Industrial Design and
Technology BA
neej88@gmail.com
Pages: 67, 191

Sunil
Patel

Industrial Design and
Technology BA
patel_sunil94@hotmail.co.uk
Pages: 99, 230

Jamie
Phillips

Product Design Engineering BSc
jamie.e.phillips@gmail.com
Pages: 126

Alina
Pîrvu

Design and Branding Strategy MA
alina.simona.pirvu@gmail.com
Pages: 174, 179

Wing Shan
Michelle Poon

Design and Branding Strategy MA
michelle_poon89@hotmail.com
Pages: 35, 162

Shilpa
Prasad

Design and Branding Strategy MA
shilpa1prasad@gmail.com
Pages: 28

Archit
Rakyan

Product Design Engineering BSc
architrakyan@googlemail.com
Pages: 127, 227

Moyo
Ralleigh

Product Design BSc
moyofujita@gmail.com
Pages: 94

Sophie
Randles

Product Design BSc
sophiekrandles@gmail.com
Pages: 98, 215

Simon
Ramm

Integrated Product Design MSc
simonramm@yahoo.co.uk
www.simonramm.com
Pages: 15, 22

Matt
Redwood

Product Design Engineering BSc
mattrewood@googlemail.com
Pages: 193

Farha
Rehnnuma

Multimedia Technology and
Design BSc
farha@misstorydesigns.com
www.misstorydesigns.com
Pages: 48

Sophie
Richards

Product Design BSc
sjrichards.design@gmail.com
Pages: 156, 223

Emily
Riggs

Product Design BSc
emilykriggs@gmail.com
Pages: 38, 100, 216

Binit
Rogunath

Mechanical Engineering BEng
Pages: 256

Shima
Roozbahani

Product Design Engineering BSc
shima.inventor@yahoo.com
Pages: 107, 131

Ziaul
Rouf

Aerospace Engineering BEng
ziaulrouf@hotmail.co.uk
Pages: 267

Amir
Ruddin

Mechanical Engineering BEng
Pages: 266

Jack
Sandys

Industrial Design and
Technology BA
jsandys.design@gmail.com
Pages: 187, 286

Devpal
Sappal

Mechanical Engineering BEng
Pages: 266

Protik
Sarkar

Aerospace Engineering BEng
sarkar.protik1@gmail.com
Pages:264, 277

Paris
Selinas

Integrated Product Design MSc
pselinas@gmail.com
Pages: 23, 32

Shazwani Farhah
Shamsuddin

Product Design BSc
shaz_pippo@yahoo.com
Pages: 235, 282

Jyoti Lakshimi
Sharma

Product Design Engineering BSc
jyotisharma@hotmail.co.uk
Pages: 58

Richard
Sheppard

Product Design Engineering BSc
sheppardesign@hotmail.co.uk
Pages: 231

Mai
Shizume

Design and Branding Strategy MA
estate-0711@tea.ocn.ne.jp
Pages: 28

Baltej Sidhu

Civil Engineering with
Sustainability MEng
Bal_sidhu@hotmail.co.uk
Pages: 30

Olly Simpson

Product Design Engineering BSc
simpson.olly@gmail.com
Pages: 56, 201

Jason Singh

Industrial Design and
Technology BA
jasonsingh89@gmail.com
Pages: 235

Roland Skinner

Industrial Design and
Technology BA
skinnerroland@googlemail.com
www.rolandskinner.net
Pages: 115, 221

Jamie Smith

Industrial Design and
Technology BA
jamiedrsmith@me.com
Pages: 233

Parisa Soltanzadeh

Integrated Product Design MSc
parisa.soltanzadeh@yahoo.com
www.idesign-product.yolasite.com
Pages: 15, 24

Dimitrios Stamatis

Industrial Design and
Technology BA
dezzzign@gmail.com
Pages: 82, 137, 212

Luke Steele

Mechanical Engineering MEng
lukejs3@hotmail.com
www.lukesprojects.weebly.com
Pages: 257

Sukumal Surichamorn

Integrated Product Design MSc
suku7th@gmail.com
Pages: 20, 32

Ozgun Tandiroglu

Integrated Product Design MSc
ozgunta@gmail.com
Pages: 32, 155

Winnie K.Y. Tang

Product Design BSc
winniekytang@gmail.com
Pages: 74

Jon Taylor

Industrial Design and
Technology BA
jonjtaylor6@gmail.com
Pages: 135, 233

Mark Taylor

Product Design Engineering BSc
mark@2partreturn.com
www.2partreturn.com
Pages: 124, 160, 225

Jamie Topp

Industrial Design and
Technology BA
toppjamie@gmail.com
www.jamietopp.com
Pages: 37,62, 208

Vasiliki Tseperka

Design and Branding Strategy MA
vassiat@hotmail.com
Pages: 174, 178

Floor Veldhuis

Design Strategy and Innovation MA
floor_veldhuis@hotmail.com
www.floorveldhuis.tk
Pages: 28, 32, 163

Phil Verheul

Industrial Design and
Technology BA
philipverheul@gmail.com
www.philv.co.uk
Pages: 37, 83, 219

Adam Wadey

Product Design Engineering BSc
aehw@hotmail.co.uk
Pages: 72, 193, 270

James Ward

Industrial Design and
Technology BA
jamesward5873@gmail.com
Pages: 26, 205

Marshel Weerakone

Civil Engineering with
Sustainability MEng
weerakone@gmail.com
Pages: 30

Beth Williams

Multimedia Technology and
Design BSc
beth_m_w@msn.com
www.beeinspired-designs.com
Pages: 52, 158

Jesse Williams

Industrial Design and
Technology BA
jlwilliams315@gmail.com
www.iamjessewilliams.com
Pages: 182, 220, 288

Alex Wilson

Integrated Product Design MSc
Pages: 15

George Williamson

Industrial Design and
Technology BA
williamsongeorge10@gmail.com
Pages: 133, 191

Deborah Wrighton

Industrial Design and
Technology BA
deborahwrighton51@gmail.com
Pages: 80, 209

Fan Wu

Design and Branding Strategy MA
wilford_wu@hotmail.com
Pages: 28

Volkan Yildirim

Civil Engineering with
Sustainability M Eng
vyildirim87@gmail.com
Pages: 30

Cheng Yu

Integrated Product Design MSc
wwq717@hotmail.com
Pages: 12

Yana Zalesskaya

Design and Branding Strategy MA
megazem@hotmail.com
Pages: 28, 33, 179

Xuefei Zheng

Integrated Product Design MSc
bernini1228@live.cn
Pages: 166

Notes

For us, Made in Brunel is something we are immensely proud to be a part of and it has fuelled our own journeys through Brunel from the first moments we knew that we were going to study here within Brunel's School of Engineering and Design.

Made in Brunel is an expanding project that is built upon year on year, it grows, evolves and its reputation gradually spreads worldwide as graduates who have been part of its portfolio build their careers. Prior to 2012 there are so many who have made this year's project possible and we could not be more thankful for the solid grounding that you left us.

Made in Brunel is run by students with the support of so many people, beyond the names mentioned within this book. We need to thank all those whose dedicated energy and time went into making Made in Brunel 2012 a success. This has been realised in many forms, from the early conceptual ideas, the long journey through creative discussions, logistical planning, project management, team building, the meetings (oh so many meetings), the hundreds of tasks from developing strategic marketing campaigns to packing thousands of invitations into thousands of little envelopes. We have loved making the journey together.

Made in Brunel itself, and the projects within this book, are built on a drive for change, a constant questioning of the world around us and the quest to find better, more responsible solutions. We are so excited to be a part of this innovation journey and hope that you find the book inspiring.

Thank you